COEVOLUTION

Trans. Ent. Soc. Lond., 1916, Pl. XII.

MODELS   MIMICS   MODELS   MIMICS

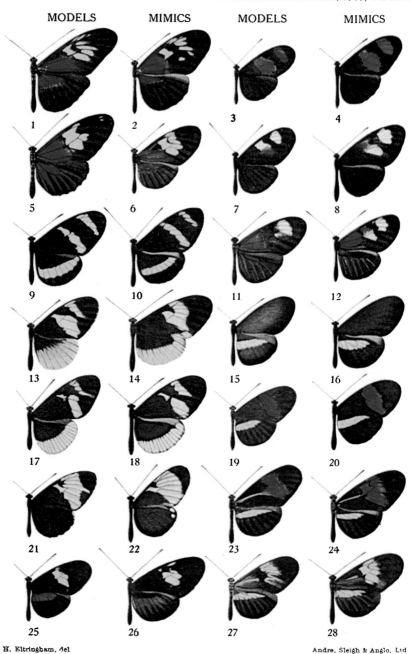

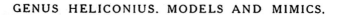

H. Eltringham, del   Andre, Sleigh & Anglo, Ltd

GENUS HELICONIUS. MODELS AND MIMICS.

COEVOLUTION

Edited by Douglas J. Futuyma

STATE UNIVERSITY OF NEW YORK,
STONY BROOK

and

Montgomery Slatkin

UNIVERSITY OF WASHINGTON

With the assistance of Bruce R. Levin
UNIVERSITY OF MASSACHUSETTS
and
Jonathan Roughgarden
STANFORD UNIVERSITY

SINAUER ASSOCIATES INC. • PUBLISHERS
Sunderland, Massachusetts   01375

THE COVER
Harry Eltringham's illustration for his paper on taxonomy and mimicry in *Heliconius* butterflies (Transactions of the Entomological Society of London, 1916). Races of *Heliconius erato* and its close relatives are the "models"; "mimics" are races of *Heliconius melpomene* and its close relatives. The question of coevolution between these groups is considered in Chapter 12. (This illustration is the frontispiece in the hardbound edition.)

Library of Congress Cataloging in Publication Data

Main entry under title:

Coevolution.

    Bibliography: p.
    Includes index.
    1. Evolution. I. Futuyma, Douglas J., 1942–
II. Slatkin, Montgomery
QH371.C73 1983   575   82-19496
ISBN 0-87893-228-3
ISBN 0-87893-229-1 (pbk.)

For information address
Sinauer Associates Inc.
Sunderland, MA 01375

Printed in U.S.A.

9 8 7 6 5 4 3 2

CONTENTS

Preface   vii

Contributors   ix

1   Introduction   1
DOUGLAS J. FUTUYMA AND MONTGOMERY SLATKIN

2   Genetic Background   14
MONTGOMERY SLATKIN

3   The Theory of Coevolution   33
JONATHAN ROUGHGARDEN

4   Phylogenetic Aspects of Coevolution   65
CHARLES MITTER AND DANIEL R. BROOKS

5   Coevolution in Bacteria and Their Viruses and Plasmids   99
BRUCE R. LEVIN AND RICHARD E. LENSKI

6   Endosymbiosis   128
LEE EHRMAN

7   Plant-Fungus Symbioses   137
JOHN A. BARRETT

8   Evolutionary Relationships between
Parasitic Helminths and Their Hosts   161
JOHN C. HOLMES

9   Parasite-Host Coevolution   186
ROBERT M. MAY AND ROY M. ANDERSON

10   Evolutionary Interactions among
Herbivorous Insects and Plants   207
DOUGLAS J. FUTUYMA

11   Dispersal of Seeds by Vertebrate Guts   232
DANIEL H. JANZEN

12 Coevolution and Mimicry   263
    LAWRENCE E. GILBERT

13 Coevolution and Pollination   282
    PETER FEINSINGER

14 Intimate Associations and Coevolution in the Sea   311
    GEERAT J. VERMEIJ

15 Coevolution and the Fossil Record   328
    STEVEN M. STANLEY, BLAIRE VAN VALKENBURGH,
    AND ROBERT S. STENECK

16 The Deer Flees, the Wolf Pursues:
    Incongruencies in Predator-Prey Coevolution   350
    ROBERT T. BAKKER

17 Coevolution between Competitors   383
    JONATHAN ROUGHGARDEN

18 Sizes of Coexisting Species   404
    DANIEL SIMBERLOFF

19 Convergent Evolution at the Community Level   431
    GORDON H. ORIANS AND ROBERT T. PAINE

Epilogue: The Study of Coevolution   459
    DOUGLAS J. FUTUYMA AND MONTGOMERY SLATKIN

    Acknowledgments   465
    Literature Cited   467
    Index   541

PREFACE

Every biologist since Darwin and Wallace has recognized that ecological interactions among species have an important influence on their evolution. Until recently there has been little attempt to develop explicit models of the evolution of these interactions. In the past several decades, however, such models have begun to emerge. The evolutionary consequences of interactions among species are now a major area of inquiry, as are the consequences of evolutionary change for the structure of ecological communities. Serving as a complement to the formal theory, evolutionary studies of ecological interactions among populations have come to constitute a major theme in field and laboratory studies as well. Studies of plants and their herbivores, pollinators, and seed dispersers, of hosts and their parasites and pathogens, of predators and their prey, and of mutualists and competitors now commonly take an evolutionary, as well as the more traditional ecological approach.

Coevolution took form as we came to realize that the principles which these different coevolving systems may have in common have hardly been explored; that the literatures of parasitology, of phytopathology, and of insect–plant ecology (for example) are read by few of the same biologists; and that much of the empirical study of ecological interactions has been only slightly influenced by the nascent theory of coevolution, just as the theoreticians have hardly begun to assimilate the vast empirical literature that bears on their theory. In soliciting the essays that make up this volume, we hoped to bring into common focus the diverse lines of study that bear on the evolution of ecological interactions, to describe the features of diverse systems of interacting species that affect their coevolution, and to identify some questions that may help to guide research in this area. We find that there are substantial differences of opinion on how coevolution should be defined, how common it is, and how it proceeds. Moreover, it is clear that a bridge between theoretical and empirical studies is in only the earliest stages of construction.

It will be evident that the development of coevolutionary theory, methods of testing such a theory, and the integration of the theory with empirical studies has barely begun. Thus, although we have attempted to facilitate exchange of ideas among the authors of these essays, they remain highly individualistic in their approach to coevolution. We have not attempted to impose homogeneity of opinion or approach on the authors and, while we attempt a brief overview of their ideas in the Epilogue, it is clear that this volume must be viewed as a first step toward synthesis rather than the synthesis we may hope will someday emerge.

We would like to express our gratitude to Mark Kirkpatrick and Steven Adolph for help with the bibliography of this volume, and to Bruce Levin and Jon Roughgarden for their contributions to its genesis. Our greatest appreciation, of course, is to the authors of these essays for sharing their ideas and knowledge.

Douglas J. Futuyma
Montgomery Slatkin

CONTRIBUTORS

ROY M. ANDERSON, Department of Zoology, Imperial College of Science and Technology, London

ROBERT T. BAKKER, Department of Earth and Planetary Sciences, The Johns Hopkins University, Baltimore

JOHN A. BARRETT, Department of Genetics, University of Liverpool, Liverpool

DANIEL R. BROOKS, Department of Zoology, University of British Columbia, Vancouver

LEE EHRMAN, Division of Natural Sciences, State University of New York, Purchase

PETER FEINSINGER, Department of Zoology, University of Florida, Gainesville

DOUGLAS J. FUTUYMA, Department of Ecology and Evolution, State University of New York, Stony Brook

LAWRENCE E. GILBERT, Department of Zoology, University of Texas, Austin

JOHN C. HOLMES, Department of Zoology, University of Alberta, Edmonton

DANIEL H. JANZEN, Department of Biology, University of Pennsylvania, Philadelphia

RICHARD E. LENSKI, Department of Zoology, University of Massachusetts, Amherst

BRUCE R. LEVIN, Department of Zoology, University of Massachusetts, Amherst

ROBERT M. MAY, Department of Biology, Princeton University, Princeton

CHARLES MITTER, Department of Entomology, University of Maryland, College Park

GORDON H. ORIANS, Institute for Environmental Studies and Department of Zoology, University of Washington, Seattle

ROBERT T. PAINE, Department of Zoology, University of Washington, Seattle

JONATHAN ROUGHGARDEN, Department of Biological Sciences, Stanford University, Stanford

DANIEL SIMBERLOFF, Department of Biological Science, Florida State University, Tallahassee

MONTGOMERY SLATKIN, Department of Zoology, University of Washington, Seattle

STEVEN M. STANLEY, Department of Earth and Planetary Sciences, The Johns Hopkins University, Baltimore

ROBERT S. STENECK, Department of Zoology and Oceanography Program, University of Maine, Darling Center, Walpole

BLAIRE VAN VALKENBURGH, Department of Earth and Planetary Sciences, The Johns Hopkins University, Baltimore

GEERAT J. VERMEIJ, Department of Zoology, University of Maryland, College Park

COEVOLUTION

CHAPTER 1

INTRODUCTION

Douglas J. Futuyma and Montgomery Slatkin

WHAT IS COEVOLUTION?

The word *coevolution* was coined by Ehrlich and Raven (1964) in their discussion of the evolutionary influences that plants and the insects that feed on plants have had on each other. Their use of the term allowed for a variety of interpretations, and it has been used differently by different authors. A restrictive definition is one provided by Janzen (1980a) and is the one adopted by many, but not all, authors in this volume. They use the term to mean that a trait of one species has evolved in response to a trait of another species, which trait itself has evolved in response to the trait in the first. This definition requires specificity—the evolution of each trait is due to the other—and reciprocity—both traits must evolve. A still more restrictive definition would also require simultaneity—both traits must evolve at the same time.

By relaxing one or more of these restrictions, we obtain definitions that other authors have used, either implicitly or explicitly. For example, mimicry among species is often regarded as an example of coevolution. But as Gilbert points out in Chapter 12, Batesian (or for that matter Müllerian) mimicry entails convergence of a mimic toward the phenotype of the model, which may not change in response to the evolution of the mimic. In this case, there is evolution due to the particular trait of the model and the behavior of the predators. Similarly, two competing species may each undergo character displacement, in which case there is clearly coevolution; but if one species diverges while the other remains unchanged, coevolution, according to the most restrictive definition, has not occurred. Yet the distinction seems quite fine and depends on probably unknowable past events.

By relaxing the criterion of specificity, the evolution of a particular trait in one or more species in response to a trait or suite of traits in several other species is included. Following Gilbert (Gilbert and Raven, 1975), Janzen (1980), and Fox (1981), we will call this "diffuse coevolution," as distinct from pairwise coevolution, which occurs in interactions between only two species. Many plants have evolved chemical and physical defenses against a diverse suite of insects, and many insects have acquired the ability to detoxify a wide range of plant chemicals. This, like the mutual adaptations of nonspecific pollinators and the flowers they pollinate, would be the result of diffuse coevolution. Yet instances can be discerned in which particular species of plants and herbivores, or plants and pollinators, have formed an intimate pairwise association and have adapted to the specific features of one or a few species. This is pairwise coevolution. Many authors who use the term *coevolution* to discuss generalized responses of groups of species to one another are discussing diffuse coevolution.

Especially when coevolution is diffuse, the criterion of simultaneity may not hold. There are many cases in which the nature of the interaction must have been similar for long periods of time, but coevolution is delayed. For example, the diverse chemical defenses of today's plants may have originated in their Cretaceous ancestors in response to herbivory by Cretaceous insects. The feeding habits of many of today's insects, however, may have evolved much more recently. Although plant evolution was guided by insect herbivory and the insect evolution was guided by the prior evolution of plant defenses, there may have been a succession of adaptive radiations widely separated in time, rather than a continual, closely coupled series of responses of particular lineages to one another. In such cases, the term *coevolution* may be used in a very broad sense, encompassing merely the adaptation of species to features of the biotic environment—features that may remain effectively constant for long periods of time.

But adaptation to an effectively constant feature of the biotic environment does not differ from adaptation to a constant feature of the abiotic environment. Coevolution, too broadly defined, becomes equivalent to evolution. Thus, of the possible definitions of coevolution, we prefer to restrict the definition and say that it has occurred when, in each of two or more ecologically interacting species, there is adaptive response to genetic change in the other(s)—a fairly narrow definition that assumes that the populations at the end of the coevolutionary process are directly descended from those at the beginning. Such a definition would exclude, for example, instances in which plants evolve defenses against beetles and the defenses are later overcome by moths while the beetles remain unchanged or become extinct.

2

However, just as the study of evolution encompasses the analysis of cases in which evolution might fail to occur, the study of coevolution encompasses cases in which reciprocal genetic responses might be expected but nevertheless do not happen. For example, without detailed knowledge of the population density and genetic characteristics of two competing species, we might expect both to undergo character displacement; if only one does, the reasons for the change in one species but not the other are part of the subject matter of coevolutionary studies. Thus, the study of coevolution is the analysis of reciprocal genetic changes that might be expected to occur in two or more ecologically interacting species and the analysis of whether the expected changes are actually realized.

HISTORY OF COEVOLUTIONARY STUDIES

Although the term is relatively new, the idea of coevolution is as old as the study of evolution itself. Darwin's discussion of pollination by insects concludes with "Thus I can understand how a flower and a bee might slowly become, either simultaneously or one after the other, modified and adapted in the most perfect manner to each other" (Darwin, 1859, p. 95). Even before 1859, plant breeders selected crops for resistance to particular parasites, especially molds and rusts (Day, 1974; see Chapter 7 by Barrett).

Some of the early evidence for Darwin's theory was provided by Bates' (1862) description of mimetic complexes. Like many of the observations used to support the theory of natural selection, mimicry had been noticed previously—oddly enough by Charles Lyell (see Lyell, 1881, pp. 417-418)—but its importance was not recognized. It is likely that Batesian mimicry usually entails specific adaptations of the mimics to the models rather than pairwise coevolution, but the possibility of changes in the model to reduce the predation caused by the presence of the mimic cannot be ruled out. Müllerian mimicry (Müller, 1878), on the other hand, may be a result of changes in one or more species in response to one another (see Chapter 12 by Gilbert) and so may be a result of reciprocal coevolution.

Almost as old as the study of coevolution are disputes about whether coevolution occurred or whether the interaction among species was established after the traits of interest evolved for other reasons. A hotly disputed case was the association of various ant species with tropical trees. There was no doubt that such associations existed and little doubt that the ants had adapted to occupy particular

3

plant species (Wheeler, 1910). What was doubted by many botanists and also by Wheeler himself was whether the plants had evolved any traits specifically for the benefit of the ants. Janzen (1966) has since provided strong evidence that traits in each species have evolved specifically to foster their mutualism. However, the general question whether it is possible to show that traits have evolved in response to particular interactions arises frequently in the study of coevolution.

Coevolution has long been recognized as a possibility in the interaction of parasites or pathogens and their hosts. As Mitter and Brooks describe in Chapter 4, systematists have long supposed that parasites and hosts have radiated in parallel. They generally discussed only in passing the question whether or not mutual adaptations of parasites and hosts became refined in the course of their association, but focused on the question whether the similarity of the parasites of different hosts could be used to determine host phylogeny, or vice versa. Mutual adaptations of hosts and pathogens, such as fungi, have been a persistent theme in plant pathology, especially in the last few decades, and numerous cases of genetic adaptations of fungi to new species of hosts have been documented (see Chapter 7 by Barrett). Probably the best-known example of coevolution in action is that of the viral disease myxomatosis introduced to rabbit populations in Australia, England, and France (Fenner and Ratcliffe, 1965). In Australia, the rabbit evolved improved resistance to the virus within a few years after the disease was introduced, and the virus evolved lower virulence, presumably because the more virulent genotypes killed their hosts before the viruses could be carried by mosquitoes to uninfected hosts. Fenner and Ratcliffe (1965, p. 345) also discuss the evolution of plague resistance by rats in India, for which geographic variation in the degree of resistance argued for past coevolution, at least on the part of the rat. In more general discussions of evolution in host–parasite systems, the emphasis has been on the conditions favoring the evolution of parasitic habits by nonparasites (see, for example, Noble and Noble, 1976), although more recently more attention has been paid to reciprocal changes by hosts (Price, 1980).

Brown and Wilson (1956) introduced the term *character displacement* to describe the result of coevolution between competing species. The general principle—that species would evolve to avoid competition by using different resources—was used by Lack (1947), who attributed the diversification of Darwin's finches in the Galápagos Islands to this principle. Using the principle of character displacement as the consequence of competition, Hutchinson (1959), and later MacArthur, Levins, Rosenzweig, Schoener, and other ecologists, developed in the 1960s and early 1970s an extensive theory of the coexistence and resource partitioning of competing species. Such morphological features as body size, head size, or bill size were proposed as measures

4

of resource use and were supposed to conform to minimal ratios among coexisting competitors. More recently, however, the ubiquity of competition and of community-level effects of competition and resource overlap has been questioned (see Chapter 18 by Simberloff) and is the subject of considerable debate at the present.

Since Ehrlich and Raven's (1964) paper, interactions among plants and animals, especially insects, have been the primary foci of the ecological literature on coevolution. The nature of chemical and physical defenses of plants against herbivory, and of insects' adaptations to these features, became a major subject in the 1970s, with contributions by Feeny, Janzen, Dethier, Gilbert, Orians, Cates, Rosenthal, and many others. The major compilation of papers on coevolution between plants and animals, edited by Gilbert and Raven (1975), drew attention to the manifold mutual adaptations of both antagonistic (e.g., plant–herbivore) and mutualistic (e.g., plant–pollinator) relationships. Many of these studies have focused on diffuse coevolution, such as the responses of plants to herbivory in general. Most of the literature concentrates on the adaptations that may have resulted from coevolution rather than on the process itself.

COEVOLUTION AS A SEPARATE FIELD

Defined broadly, coevolution encompasses much of evolution and can hardly lay claim to any distinction as a subject of study. However, progress in understanding particular areas within evolution is often enhanced by drawing attention to them as definable subjects, as the brief history of sociobiology illustrates. Coevolution calls for special attention primarily because it is a major point of contact between evolution and ecology. Evolutionists have long argued that biotic interactions have fostered the evolution of adaptations. Biotic interactions such as competition may prevent a group from radiating or, by providing novel opportunities, may trigger a radiation. Competition among members of such groups as the African cichlids and the Hawaiian honeycreepers may promote adaptive diversity and specialization. Morphological and behavioral innovations may evolve in response to parasites, predators or symbionts. Stebbins (1974) has argued that all of these factors played important roles in the evolution of flowers, fruits, and seeds. Not all adaptations are responses to interactions among species, but many evolutionary events must be understandable only in the context of those interactions.

Evolutionary ecology includes the study of interspecific interactions, and many ecological phenomena are usually thought to be ex-

5

plicable by evolutionary principles. Levins (1968) predicted that the diversity of competing species is controlled by the evolution of characters permitting resource partitioning. MacArthur (1972) concluded that the food webs in species-rich and species-poor biotas are controlled by the evolution of specializations. Ehrlich and Raven (1964) explained the trophic specializations of insects as responses to the chemical features of their host plants. Slobodkin and Sanders (1969) and Southwood (1961) described how evolution will lead to increased numbers and diversity of species in communities. May (1973) argued that communities comprising coadapted or coevolved species tend to be more stable. The idea that coevolution, as well as the impact of immigration and extinction, should lead to predictable ecological relationships among species and to a predictable community structure has led to the question whether independently evolved communities should converge in structure (see Orians and Solbrig, 1978; Cody, 1974; Chapter 19 by Orians and Paine). If community ecology is to be built on an evolutionary theory of species interactions, an understanding of coevolutionary processes is essential.

The study of coevolution forces a different view of genetic evolution than is usually adopted. In population genetics and evolutionary theory, each species is usually considered in isolation, with the environment and associated species relegated to the background, which is assumed to remain unchanged. Coevolutionary theory, as discussed by Roughgarden in Chapter 3, assumes that genetic changes may occur in all interacting species, allowing genetic changes to be driven both by immediate interactions and by the feedback through the rest of the community. The distinctive feature of coevolution is that the selective factor (e.g., a predator) that stimulates evolution in one species (e.g., a prey) is itself responsive to that evolution, and the response should be predictable. In some cases a coevolutionary equilibrium may be established. In other cases there may be no coevolutionary equilibrium, and evolution may continue over longer time scales than are typical for the attainment of gene frequency equilibria as usually treated in population genetic models. For example, as new mutations arise, a prey or host species could evolve new defenses over a long period while its principal predator or parasite evolves new ways to overcome those defenses. This "arms race," as several authors (e.g., Gilbert, 1971; Dawkins and Krebs, 1979) have described it, requires that the rate of origin and the phenotypic effect of new mutations enter the models.

The study of coevolution also requires a different view of the time course of evolution. If one species is considered alone, it would be expected to evolve until it has met whatever challenges it faced and then to stop. For a variety of reasons discussed in Chapter 2, natural selection in a particular direction does not produce continued progress. But

6

if two or more species are evolving in response to one another, then continued progress in each species might occur. A prey or host species could continually produce new defenses while its principal predator or parasite finds new ways to overcome those defenses.

The intensity of ecological interactions can often be measured. If so, then it is possible to estimate the potential strengths of selection in the interacting species. If it is found that species have not coevolved despite pressure to do so, that would call for an explanation. Species might not have been associated for long enough or there might have been insufficient mutations of the appropriate kind. If coevolution turns out to be a rare or sporadic phenomenon, we must ask whether species interactions have been weak or of short duration or whether evolution is strongly constrained by lack of mutations.

Attention to coevolution could raise and help provide answers to many questions about the history of evolution. How often has the adaptive radiation of a group been dependent on the radiation of other groups with which they interact? Does the speciation of hosts and parasites often occur in parallel? Do defense systems of prey become more complex over evolutionary time with the addition of new defenses to the armamentarium, or are old defenses traded for new ones? Do parasites tend toward specialization or toward a benign or even mutualistic relationship with their hosts? In general, how much of the history of evolution must be explained in terms of the evolutionary effects of interspecific interactions?

Finally, the study of some coevolved systems can have direct or indirect practical benefits for agriculture and medicine. If we can understand why insects and pathogens have sometimes been able and sometimes unable to overcome the defenses of plants, we may be able to judge whether the breeding of resistant crops is doomed to failure because of counterevolution by their pests. It would be useful to know whether plant resistance is attained by single defensive features or by complexes of defenses and to know whether pest control can be best achieved by planting monocultures or intercropped polycultures of resistant strains. In medicine, we should like to know to what extent pathogens and parasites are genetically variable for virulence and to what extent populations long exposed to a particular pathogen are more resistant than unexposed populations. What can be learned about the mechanisms of resistance from comparing susceptible and resistant populations or from cross-infection studies of host-specific species? The experiments required to elucidate coevolutionary questions, such as the reasons for host specificity of parasites, could lead to advances on practical fronts.

APPROACHES TO THE STUDY OF COEVOLUTION

In principle, it is rather easy to demonstrate the adaptation of one species to particular features of another. The attraction of crucifer-feeding flea beetles to the allyl isothiocyanate of their host plants, the mimicry of the monarch butterfly by the viceroy, or the venom apparatus of a rattlesnake that subdues its prey are easily seen to be adaptations to particular traits in other species. The demonstration of pairwise coevolution is more difficult, however, because it must be shown that two or more species evolved in response to one another. And diffuse coevolution may consist of events widely separated in evolutionary time and may involve the adaptation of one species to a class of species whose features evolved in ancestors long gone. For example, cyanogenic glycosides of a cherry tree, which tent caterpillars and other insects can detoxify with the enzyme rhodanese (Chapter 10), presumably evolved in an early rosaceous ancestor in response to a wide gamut of insects. It would be difficult to demonstrate or even argue convincingly that cherries evolved cyanogenic glycosides specifically in response to tent caterpillars and that tent caterpillars evolved rhodanese specifically in response to cherries. One could hope to show only that cyanogenic glycosides are rosaceous adaptations to herbivory and that rhodanese is a lepidopteran adaptation to plant toxins or to the cyanogenic glycosides that many plants possess.

It is most likely that evidence can be found to demonstrate pairwise coevolution, the mutual adjustments of two or a few closely associated species to each other. There are several lines of inquiry that could provide evidence of reciprocal coevolution.

Direct observation of genetic changes

An excellent example of the observation of a direct response to interactions is the case of the myxoma virus and rabbits mentioned already (Fenner and Ratcliffe, 1965). Such cases are rare, but a similar one is the appearance of strains of the Hessian fly (*Mayetiola destructor*) that are able to attack a series of sequentially planted resistant strains of wheat (Gallun, 1977). In this case the wheat "evolved" according to a controlled plan, rather than as a natural response to the evolution of the fly. Barrett (Chapter 7) describes similar cases of genetic changes in plants and pathogens.

Laboratory model systems of the coevolution of competitors have been studied by Park and Lloyd (1955) in *Tribolium*, by Moore (1952), Seaton and Antonovics (1967), Futuyma (1970), and others in *Drosophila*, by Pimentel et al. (1965) in muscid flies, and by Chao et al. (1977) in bacteria-bacteriophage interactions (see also Chapter 5 by Levin and Lenski). These experiments provide evidence that rapid coevolution can occur in some cases but not others. Pimentel and his co-workers (e.g., Pimentel and Stone, 1968) have found evidence of genetic changes in both houseflies (*Musca*) and the parasitoid wasp *Nasonia vitripennis* when cultured together, and Hassell and Huffaker (1969) reported increased resistance in the host and increased effectiveness of the parasitoid in a moth–wasp laboratory system. Such studies show, of course, that pairwise coevolution is possible, not that it commonly occurs in nature. In the absence of an actual history of the dynamics of genetic change, the demonstration that each of two interacting species is genetically variable for the characteristics that affect their interaction can at least show the potential for coevolution.

Fossil evidence

The fossil record can provide evidence of coevolution on a vastly longer time scale than is accessible in laboratory and experimental studies. The ideal paleontological evidence would be a continuous deposit of strata in which each of two species shows gradual change in characters that reflect their interaction. For example, Kellogg (1980) provides evidence for character displacement of size in fossil radiolarians. However, the fossil record is seldom complete enough to provide much detailed information, and coevolution is more often inferred than demonstrated. For example, the evolution of hypsodonty in horses is often cited as a coevolutionary response to increased silica in grasses, but we are not aware of evidence that grasses steadily evolved a higher silica content as the horses were evolving hypsodonty. Although the fossil record can show evidence of coevolution only on a geological time scale, it provides the only way to assess the macroevolutionary importance of coevolution.

Taxonomic evidence

Very strong evidence for coevolution of two groups is obtained if the phylogenetic trees of the two groups are congruent or nearly so. We will call this parallel cladogenesis, which does not imply that coevolution is responsible for the diversity of either group, only that the two groups have remained associated with each other. Parallel cladogenesis might be expected for parasites and hosts and has often been claimed in this context. The evidence for parallel cladogenesis (see Chapter 4 by Mitter and Brooks) appears to be clear in some groups and much less so in others.

Taxonomic evidence for the coevolution of adaptive characters need not entail cladogenetic congruence of large taxa. For example, phylogenetic analysis may establish that ancestral and derived states of a defensive character in a host species might be associated with ancestral and derived characters that facilitate exploitation of the host by parasites whose relatives occupy unrelated hosts. In fact, the independence of taxonomic and adaptive patterns may provide strong evidence for the adaptive significance of particular characters.

Functional morphology and ethology

Careful analysis of unusual or unique characters of each of two interacting species can show these features are one species' adaptations to the other species. This evidence is indirect and often not conclusive but may be the only kind available. For example, tannins and other compounds in plants are often cited as defenses against insects, but some authors have raised the possibility that they are primarily defenses against pathogens, and others have questioned whether they evolved for defensive reasons at all—even if they now serve in that role. In other cases, characters appear to be so exclusively designed to mediate ecological interactions that it is hard to imagine any other reason for their evolution. For example, despite Wheeler's (1910) doubts about the evolution of acacias and their associated ants, the proteinaceous bodies on the leaves, on which the ants feed, are restricted entirely to species of acacias that harbor *Pseudomyrmex* ants that live nowhere else. Similarly, the evidence of coevolution is very strong in cases such as aphids that have lost the ability to clean themselves of honeydew, which develops mold if not removed by ants.

Many examples of mutualisms can be understood only as the result of reciprocal coevolution, especially when the interaction entails a very exclusive relationship between the species. Pollinators and the flowers they pollinate show some of the most striking mutual adaptations, but as Feinsinger (Chapter 13) discusses, it is sometimes difficult to argue for the coevolutionary origin of those traits. Janzen (Chapter 11) discusses similar problems that arise for fruits and their dispersal agents.

For hosts and various kinds of parasites, Ehrman (Chapter 6), Barrett (Chapter 7), Holmes (Chapter 8), May and Anderson (Chapter 9), and Futuyma (Chapter 10) find abundant evidence of the adaptation of parasites to specific features of their hosts. The extreme specialization of many parasites to one host or to a few closely related hosts argues for their long-term association. There is less evidence of specific adaptations of hosts to particular parasites. Instead, more generalized defenses against suites of parasites seem to be much more common, as would be expected if hosts were attacked simultaneously by several species of parasites. However, there are numerous exceptions, such as

10

the evolution of resistance by rabbits to myxomatosis, discussed earlier, and the adaptation of *Drosophila* to certain endosymbionts (see Chapter 6 by Ehrman).

Geographical patterns

Another common source of evidence is geographical patterns, exemplified by the study of Brown and Wilson (1956). The strongest evidence comes from the comparison of allopatric populations of each species with sympatric populations. The usual approach is to directly compare the behavior and morphology of species in allopatry and sympatry, but, more recently, Hairston (1980) has shown the power of experimental methods for making the same comparison. Hairston found that where two species of salamanders have very narrow altitudinal overlap, suggesting the operation of strong competition, removal of one species was followed by a pronounced increase in the density of the other. In another mountain range, the altitudinal overlap is much greater, which suggested that the species had diverged in their use of resources. The removal of one species in this zone of overlap had a much smaller effect on the density of the other.

Parallel geographical variation of two broadly sympatric, interacting species may show the adaptive response of one species to another but will seldom be adequate to demonstrate pairwise coevolution (see Chapter 14 by Vermeij). The Müllerian mimic butterflies, *Heliconius erato* and *H. melpomene*, for example, display extraordinary parallel variation (Turner, 1976), but this in itself will not reveal whether one has converged on the other or both to each other (see Chapter 12 by Gilbert). In some cases, a strong plausibility argument can be based on the specificity of the interaction. For example, if the tongue length of a host-specific bee varies in parallel with the corolla length of its host, but the host plant is serviced by many species of pollinators, it is likely that the bee has adapted to the plant but not vice versa (see Chapter 13 by Feinsinger).

If, for a predator–prey or host–parasite interaction, we know only that parasites from each of several localities can exploit only their own host populations, we cannot determine whether coevolution has been reciprocal or unilateral. If parasite populations a and b perform better on their respective host populations A and B than in reciprocal tests, we do not know whether there has been adaptation of the parasites to hosts that are geographically variable for some other reasons or reciprocal evolution of resistance by the host that is insufficient to compensate for the parasite's evolution. It is necessary to perform cross-infections with both "naive" allopatric and "experienced" sympatric populations of both host and parasite. Ideally, one could show, for example, the resistance to malaria of sickle-cell heterozygotes, demonstrate that they are more prevalent in malarial regions, and find that *Plasmodium* from regions of high sickle-cell frequency reproduce more effectively in sickle-cell heterozygotes than do *Plasmodium* from regions of low sickle-cell frequency. There is a paucity of such demonstrations of reciprocal coevolution, although they may not be too difficult to perform.

Theoretical methods

Most of the approaches taken to the study of coevolution, including most of those taken in this book, are empirical. But theory can also play a role by making predictions that can be tested with further empirical studies. Levin and Lenski (Chapter 5), May and Anderson (Chapter 9), and Roughgarden (Chapter 17) have all taken that approach. In all these cases, mathematical models show how the observed patterns can be accounted for by coevolutionary processes. The agreement with a coevolutionary model does not prove the occurrence of coevolution, but it can often add support to the hypothesis that coevolution has occurred and can suggest other experiments or observations that could provide further support.

Although coevolutionary theory, as reviewed by Roughgarden (1979; and Chapter 3) and by Slatkin and Maynard Smith (1980), has played an important role in some coevolutionary studies, there is still a large and obvious gap between the theoretical and empirical aspects of the subject. The reason for this gap is partly that, in many systems, the principal question is whether or not coevolution has occurred, whereas the mathematical models are directed to showing how coevolution will occur under the assumption that it will do so. This gap will be partly closed if field and experimental workers can address the assumptions and conclusions of some of the models and if theoreticians can develop their models with more of an eye for their empirical utility.

Community-level analysis

The approaches described so far concentrate on interactions between pairs of species and the consequences of those interactions. An alternative approach is to look for community-level patterns that might be caused by coevolution. The failure to find such patterns would suggest that coevolution may not be an important process in structuring communities even if some of the component species have coevolved with one another. This conclusion is similar to that reached in some of the discussions of coevolutionary patterns in the fossil record. Although coevolution might have been occurring, it may not be manifest at all levels.

Orians and Paine (Chapter 19) examine the evidence of community level convergence in a variety of marine and terrestrial systems. Simberloff (Chapter 18) describes several recent studies by him and his co-workers on the statistical analysis of community-level patterns. Simberloff's approach has been to generate null models—models generating patterns expected if interactions were not important—and test observed patterns for significant departures from the null models. Although this approach is controversial, the controversy itself will stimulate more detailed studies and more careful interpretation of data.

IN CONCLUSION

The scope of coevolutionary studies, like that of the study of evolution generally, is very broad, ranging from the intricate coadaptation of intimate associations that include even intracellular symbionts to the manifold interactions among species that may play a role in shaping community structure. The chapters in this volume address a wide spectrum of subjects. The authors, who include systematists, paleontologists, ecologists, and geneticists, take different approaches to the study of coevolution and often reach different conclusions. As editors, we have made no attempt to disguise the differences among the various authors or to forge a consensus on the prevalence or importance of coevolution. Our goal was not to produce a synthesis of coevolutionary studies; the field is far too new and diverse for that to be successful. Instead, we have tried to show the breadth and richness of the subject, which may not be apparent from coevolutionary studies of only one group of organisms or of only one type of ecological interaction.

CHAPTER 2

GENETIC BACKGROUND

Montgomery Slatkin

The present state of knowledge about the genetic basis of phenotypic evolution is not complete enough to be summarized by a few simple principles that can be applied in coevolutionary studies. Instead, several factors are thought to govern genetic evolution, and different views of their relative importance are held. In any coevolutionary study, it is usually necessary to adopt one or more of these views, and even when assumptions about genetic evolution are not made explicit, some assumptions are necessarily made.

In this chapter I will discuss the different views of genetic evolution that are most relevant to coevolution. I will generally not advocate one view over another but will emphasize the implications for a coevolutionary study of adopting one or another view. Because coevolution—however it is defined—is the consequence of ecological interactions among species, I will focus on the genetics of phenotypic responses to selection, first by reviewing the theory of selection at different levels, then by examining the different views of the maintenance of genetic variation, and finally by discussing the many factors that can act to constrain evolutionary responses to selection.

LEVELS OF SELECTION

The principal mechanism of Darwin's theory of evolution is natural selection acting on individual differences. As Lewontin (1970) has pointed out, however, any group of objects, from molecules to galaxies, will evolve if they possess heritable differences that are correlated with differences in survival and reproduction. It is not only possible but inevitable for evolution to occur when this condition is met. Evolution can result from selection of individual differences but also from selection of groups of related individuals, populations, species, and entire communities. There are currently strong disagreements about the relative importance of selection at different levels. The population genetic theories of the 1920s and 1930s, which constituted the "modern synthesis" (Huxley, 1942) and which completed the integration of Mendelian genetics into evolutionary theory, focused on individual selection, but many ecologists and ethologists assumed that selection at the level of the population or the species was also important. In the development of sociobiology, selection among kin groups has been invoked to account for many altruistic and social behaviors. The debate over macroevolutionary patterns and their explanation partly depends on the importance of selection among species. Although G. C. Williams (1966) advocated the "principle of parsimony" in evolution—if any adaptation could be explained through individual selection, selection at other levels should not be invoked—a more pluralistic view of levels of selection has emerged. Any coevolutionary study must make some assumption about the level at which selection is operating.

Individual selection

Individual selection is due to differences among individuals in survival and reproduction, which together determine an individual's fitness. An individual's fitness may depend on population size (density dependence) or the distribution of other phenotypes in the population (frequency dependence). In fact, many kinds of ecological interactions probably lead to density-dependent or frequency-dependent fitnesses.

The phenotypic changes due to individual selection depend on both the amount of variation in a population and the intensity of selection. Fisher (1930) formalized this idea and put it in genetic terms with what he described as "the fundamental theorem of natural selection," which states that the rate of evolution of a character at any time is proportional to its additive genetic variance, which is the heritable component of a character's genetic basis. The constant of proportionality is the quantitative measure of the intensity of selection. Intuitively, Fisher's theorem implies that individual selection will tend to make a species better adapted to environmental conditions. Fisher's theorem is exactly true only under rather special conditions and is not true when there are density- or frequency-dependent interactions, linkage among loci, or correlations among characters caused by pleiotropic effects of genes (Roughgarden, 1979; Ewens, 1979). But there are many other cases in which Fisher's theorem is approximately true in the sense that individual selection nearly always leads to an increase in the average fitness of a population (Ewens, 1979). Ecological interactions

15

among members of the same species and of different species would tend to lead to fitnesses that depend partly on total population size. Roughgarden (1971), Charlesworth (1971), and Anderson (1971) have shown that density-dependent natural selection will tend to maximize the total population size of a population. This result is analogous to Fisher's theorem.

One view of the role of individual selection in evolution is that of Darwin and, in genetic terms, of Fisher. This view is that much of evolution consists of the gradual and steady improvement of different characters under the action of natural selection. A character stops evolving when it cannot be improved upon under current ecological conditions. Individual selection is, in this view, both creative and stabilizing.

This view of evolution can be illustrated by an "adaptive topography" (Wright, 1931). As shown in Figure 1, fitnesses of individuals with different measures of two phenotypic characters can be represented by a surface, with the two horizontal dimensions being the ranges of possible values of the characters and the vertical dimension being fitness. The environmental conditions determine the actual shape of the surface, and the surface in turn determines the course of evolution of those two characters. According to Fisher's theorem, evolution will proceed uphill on the surface and stop when a peak is reached.

If there are two or more peaks in the adaptive surface, the peak that a species ultimately reaches is determined by its initial phenotypic composition. In Figure 1, species A, which starts with small values of the two characters, will evolve to peak I, whereas species B, which initially has larger values of the characters, evolves to peak II. Unless environmental conditions change, thereby altering the shape of the adaptive surfaces, Fisher's theorem predicts that the two species will not evolve once they reach their respective peaks, which represent evolutionarily stable combinations of the two characters. It is especially important that species A, at the lower peak, will not evolve to occupy peak II, the higher peak. Even though there is a superior combination of characters, species A cannot attain that combination, according to Fisher's theorem, because it cannot cross the "adaptive valley" that separates the two peaks.

With this view of the role of individual selection, the ecological conditions, including interactions with other species, are the driving forces of evolution. When conditions change (as could occur when other species in the community evolve), the adaptive surface changes, causing the species to evolve to a new peak. Almost by necessity this view is the one usually adopted in coevolutionary studies, because the emphasis in coevolution is on interactions among species.

Although there is no real dispute with Fisher's view of how in-

16

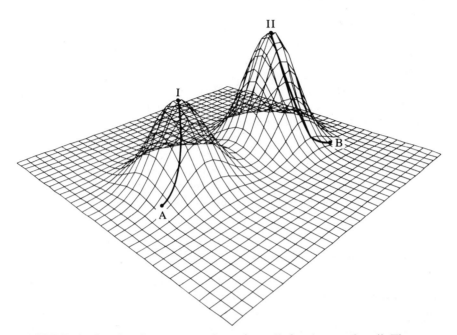

FIGURE 1. Graphical representation of an "adaptive surface." The two horizontal axes represent two phenotypic characters, such as body size and color, and the vertical axis represents the mean fitness of a population that has the particular average values for the two traits. The two peaks, I and II, represent two combinations of these characters that are particularly well coordinated. According to Wright's and Fisher's genetic theories, species will tend to evolve "uphill" on such surfaces. Species A will tend to reach an evolutionary equilibrium at peak I and species B at peak II. Despite the fact that peak II is higher than peak I, indicating that one combination of characters is better than the other, species A cannot evolve to peak II without some other genetic mechanism like genetic drift acting in addition to individual selection.

dividual selection causes genetic changes, alternative views of the role of individual selection arise from different assumptions about how environmental changes affect the adaptive topography. Fisher assumed that evolutionary changes follow environmental changes that alter the adaptive topography. Wright (1931), in contrast, argued that evolutionary changes occurred when a species moved to another adaptive peak through the action of other forces, particularly genetic drift and group selection. In Wright's view, individual selection is more of a conservative force, with evolutionary changes not necessarily being driven by environmental change. Chance events also play an important role.

17

Wright's view of individual selection is not incompatible with Fisher's. They both emphasize the power of individual selection, but Wright sees the need for other mechanisms of genetic changes as well. Wright's view has been adopted by the proponents of the punctuated equilibrium theory of evolution (Eldredge and Gould, 1972; Stanley, 1979). Part of that theory is that individual selection is relatively ineffective in causing phenotypic changes in widespread species. The punctuational theory differs from Wright's because it assumes that phenotypic changes, at least those observable in the fossil record, are associated with species formation rather than phyletic evolution. In the punctuational theory, an entire species does not move from one adaptive peak to another. Instead a species gives rise to a new species which might occupy another adaptive peak. But the punctuationalists agree with Wright that individual selection may play a more conservative and less creative role in evolution than in Darwin's and Fisher's views.

Kin selection and group selection

Kin selection is the term used to describe selection based on differences among collections of related individuals (Maynard Smith, 1964), and *group selection* is the term used to describe selection due to differences among populations of a species (Wynne-Edwards, 1962; Maynard Smith, 1964). Both Fisher (1930) and Haldane (1932) were aware that genetic evolution could occur through effects of an individual on its relatives. Fisher (1930) suggested that the evolution of aposematic coloration would be due to the increased protection of an individual's relatives gained by a predator's learning to avoid a distasteful prey. Parental care, although technically due to kin selection, was regarded as being so obviously adaptive to the parents that it did not require a separate explanation. It was not, however, until the formal development of a theory of kin selection by W. D. Hamilton (1964a,b) and the demonstration that kin selection theory could explain a variety of social phenomena—especially the evolution of eusociality in Hymenoptera—that the potential power of kin selection was recognized.

Group selection was first discussed in genetic terms by Wright (1945), who called it *interdemic selection*. He noted that genetic evolution could occur if there were genetic differences among local populations that caused differences in extinction and migration rates. Lewontin (1962) argued that group selection could be responsible for the observed low frequencies of the t allele in *Mus musculus*. Because of meiotic drive, the t allele increased in frequency in each population even though the t/t individuals were sterile. Levin, Petras, and Rasmussen (1969) used a computer simulation to show that local extinction of populations caused by fixation of the t allele could account for its low frequency, albeit under rather restrictive assumptions about deme sizes and migration rates.

Many ethologists implicitly assumed the operation of group selection to explain the presence of behaviors that seem to confer no advantage to the individual. For example, Huxley (1966) argued that elaborate display rituals evolved to settle contests among individuals because physical combat would be bad for the entire social group. Wynne-Edwards (1962) argued that many behaviors evolved through group selection as a mechanism to allow individuals to assess their local population densities and limit their reproduction to prevent overexploitation of resources. G. C. Williams (1966) countered Wynne-Edward's views with the argument that group selection was not needed to explain the presence of most social behaviors. Williams suggested a "principle of parsimony" to be used in evolutionary discussions: if individual selection could account for an adaptation, then there was no reason to invoke group selection. This principle assumes that individual selection is more parsimonious than group selection or selection at other levels.

Numerous theoretical studies have shown that group selection is theoretically possible when populations are extensively subdivided and local extinction is reasonably common (Wade, 1978; Roughgarden, 1979), and Wade (1977) has demonstrated that substantial phenotypic evolution can result from group selection in laboratory populations of *Tribolium*. But what is not known is whether conditions often exist for group selection to produce substantial evolution. Wilson (1980) and Price (1980) contend that many species have population structures conducive to group selection. Parasites in particular have highly subdivided populations and local populations can become extinct by killing their hosts. Group selection is often assumed to be the explanation for the loss of virulence of some disease parasites like the myxomatosis virus in Australian rabbits (Fenner and Ratcliffe, 1965).

Species selection

The term *species selection* was introduced by Stanley (1975), although Fisher (1930), Lewontin (1970), and Eldredge and Gould (1972) all discussed the idea. Species selection can be effective if there are differences among species in extinction and speciation rates. Any character correlated with either or both of these rates would increase in frequency in a group of species considered together. There would tend to be more species with characteristics causing high speciation rates and low extinction rates. For example, it is possible that monophagous insect

19

species have greater tendencies to form new species than do poly-phagous species because of greater opportunities for geographical iso-lation. If species derived from monophagous species are also mono-phagous, then we could expect the species characteristic monophagy to become common purely by species selection. On the other hand, mo-nophagy could tend to increase the extinction rates of species because of the commitment to a single resource, thereby causing monophagous species to become less frequent.

This point is particularly relevant for coevolutionary discussions that rely on comparisons among species. To continue the preceding ex-ample, suppose it were found that there were more monophagous species associated with particular kinds of trees—say, those with large amounts of alkaloids. If only individual selection were considered, the conclusion would be that the presence of alkaloids caused the evolu-tion of monophagy in species associated with those trees. An alter-native explanation is that the alkaloids reduced the speciation rates or increased the extinction rates of the polyphagous species associated with those trees. With that explanation, a polyphagous species feed-ing on alkaloid-rich trees would not necessarily evolve to become monophagous.

Most coevolutionary studies focus on individual selection rather than selection at other levels. If the view of the punctuationalists is correct and species selection has been an important creative force in macroevolution, then, as Stanley (1979) has discussed, many coevolu-tionary studies will have to be reinterpreted. On the other hand, coevolutionary studies, including those described in this book, may provide evidence for the efficacy of individual selection in established species and have some bearing on macroevolutionary theory.

Ecosystem selection

Associations among species in ecosystems could cause each species to have lower extinction rates and higher dispersal rates. The result could be the evolution of associations of species that are mutually beneficial due to the selection of entire ecological communities. This idea has been discussed at various times by ecologists, some of whom have regarded ecological communities as "superorganisms" made up of mutually coadapted species (Allee et al., 1949). Wilson (1980) has more recently generated new interest in the possible importance of selection among ecosystems.

Wilson's (1980) argument is that there are too many features of species in communities that work to the advantage of other species to be explained purely through the action on individual selection in each species. Individual selection (and kin and group selection as well) de-pend on the genetic gain as measured by gene frequencies within the species. A species would not work in the interest of another species, even if that results in the eventual extinction of other species and ultimately itself. For example (Wilson, 1980, p. 5): when earthworms restructure the soil, plant growth is promoted and the long-term availability of food is thereby ensured. However, individual selection would favor traits that make worms more efficient at feeding despite any eventual damage to the plants. This effect of individual selection could be offset by selection among ecosystems: those ecosystems containing earthworms that benefit plants would survive for longer times and would tend to produce colonists for other ecosystems.

The consideration of selection at the level of ecosystems raises two separate questions. One is whether the apparent harmony among species is in conflict with the theory of individual selection. Wilson (1980, Chapter 5) reviews several examples where that may be the case, but in each example there are alternative explanations possible. The second question is whether the observations require selection among ecosystems, as proposed by Wilson (1980) or whether they can be explained by coevolution within the community. As discussed by Roughgarden (Chapter 3, this volume), genetic changes in a species can be affected by the "feedback" through interactions with other species in the community. According to this theory, a species like the earthworm would not be selected to destroy its food supply because the reduced food supply would tend to oppose that selection.

THE PARADOX OF VARIATION

The response to selection at any level depends on the existence of heritable variation among the units being selected. But one of the consequences of most kinds of selection is reduction in the extent of variation. Any view of genetic evolution as primarily due to selection at one level must contain assumptions about how variation at that level is created or maintained.

There is abundant genetic variation among individuals in most populations. Artificial selection of most phenotypic characters in most species readily produces significant changes. The implication for coevolution is that if selection due to ecological interactions is strong enough, then the trait would be expected to evolve quickly. At the genetic level, biochemical studies have shown that numerous loci have more than one allelic state (Lewontin, 1974). The principal question is whether genetic variation among individuals is in itself adaptive or not.

Lewontin (1974) distinguishes two schools of thought about the

21

maintenance of genetic variation. One—the "classical" school—views most heritable variation at any time as being not adaptive. The role of individual selection on most characters is to eliminate deleterious mutations. The genetic model regarded as typical by members of this school is of a locus with mutations to deleterious alleles being balanced by selection against those alleles. When environmental conditions change, alleles that were deleterious might become advantageous, thereby adapting a population to the new conditions. The other school—the "balanced" school—views most genetic variation as being adaptive and maintained in species by individual selection alone. The genetic model viewed as being typical by this school is a locus with two or more overdominant alleles, for which the heterozygotes have higher fitnesses than the homozygotes. When environmental conditions change, a different genetic balance may be reached, but selection would still act to preserve genetic variation.

The extreme views attributed to the two schools are not held by many evolutionary geneticists. Most would adopt a position intermediate between these two extremes. But the differences between the views are important, particularly for studies of coevolution. If the view of the balanced school is adopted for a particular character or set of characters, then it is always reasonable to assume that sufficient genetic variation exists to enable each species to respond immediately to new conditions, possibly to conditions created by genetic changes in other species. If the view of the classical school is adopted, then there could be some delay in a species' response to new conditions because the appropriate mutations may not be present. One way to understand the difference between the two schools is to imagine what would happen if the mutation process were somehow stopped. The classical school would predict that genetic evolution would stop quickly and there would be very little potential for response to any new conditions. The balanced school would predict that genetic evolution would continue for some time before stopping. Each species would have a store of genetic variants, and evolution would proceed until that store was exhausted.

There is at the present time no way to determine which view of genetic evolution is more nearly correct. It was thought that the application of biochemical techniques would resolve the issue when it was shown that there was abundant genetic variation at the molecular level (Lewontin, 1974). But, instead, the issue in population genetics became the question whether the observed genetic variation was selectively neutral. For the study of coevolution, the neutral mutation theory of Kimura (1968) and King and Jukes (1969) is not central. But the extent to which phenotypic evolution is limited by a lack of genetic variation at any time is. Coevolution is concerned with evolutionary responses to ecological conditions that can change quickly. If evolutionary changes cannot occur on an ecological time scale, coevolution may not be found even when expected.

Selection at higher levels also depends on variation. Laboratory experiments (Wade and McCauley, 1980) and theoretical studies (Eshel, 1972; Slatkin, 1977, 1981b) showed that genetic drift can act like mutation in creating random differences among local populations. Uyenoyama (1979) showed that random fluctuations in selection intensity in different populations can also lead to local genetic differentiation. But there are no field studies showing that there are characteristics of local populations that are both heritable and also subject to group selection. The extremely subdivided populations of many species, especially parasitic species, provides circumstantial evidence that group selection could be important (Wilson, 1980; Price, 1980), but the critical field studies remain to be done.

For species selection to be a creative force, there must be some mechanism creating differences among species. Stanley (1979), Gould and Eldredge (1977), and Vrba (1980) argue that the fossil record shows that large and unpredictable phenotypic changes are likely to occur when new species are formed. For example, if a new species is formed from a species with a certain average body size, the new species may differ from its parent species in average body size; but it may be larger or smaller. The apparent unpredictability of the changes occurring at speciation do not necessarily imply that speciation is due to random events such as genetic drift (Stanley, 1979; Charlesworth, Lande and Slatkin, 1982). But they may appear random when a large number of speciation events are considered. Stanley (1979) makes the analogy that speciation could act like the mutation process within a population. It could produce variation on which species selection could later act.

CONSTRAINTS ON GENETIC EVOLUTION

Natural selection cannot improve a species indefinitely. Eventually and possibly very quickly, evolution in a particular direction will stop because of one or more constraints. There can be several factors, which are not mutually exclusive, that contribute to limiting the genetic response to selection. All of these factors are important for some species, but there is not wide agreement about their overall importance in evolution because it is impossible to say what is typical of most species or even most species in one taxon.

Pleiotropy

Pleiotropic genes directly affect two or more characters and are commonly found in genetic studies. The *yellow* allele in *Drosophila melanogaster* causes the eye to be yellow and also causes the body to be yellow and modifies the shape of the spermatheca (Dobzhansky, 1970). Dobzhansky (1970) and Wright (1967) have long emphasized the importance of pleiotropy to evolution. Pleiotropy can constrain the evolution of one character because of changes in other characters affected by the same genes. Even if, somehow, environmental conditions favored *Drosophila* with yellow eyes, changes in the spermatheca could cause such a large decrease in reproductive success that the *yellow* allele could not increase in frequency.

At the phenotypic level, pleiotropy is manifested as correlations among characters. Selection on one character results in changes in others. The correlations among characters due to pleiotropy can evolve through modifier genes that alter the pleiotropic effects of other genes or through the increase in frequency of genes that control one of the characters alone. For example, yellow eyes could evolve in *D. melanogaster* if an allele at another locus reduced the effect of the *yellow* allele on the spermatheca or if there were an allele at another locus that produced yellow eyes but did not affect the spermatheca.

Pleiotropy is thought to be an important cause for a phenonenon commonly found in populations subject to strong directional selection. A quantitative character that is strongly selected usually responds quickly for the first few generations. Then a "plateau" is reached after which further progress cannot be achieved. In some cases, no further progress can be made because the selected lines become sterile. The level of the plateau is difficult to predict. Waddington (1960) selected populations of *Drosophila melanogaster* for salinity tolerance, producing populations that could survive on a medium containing 7% salt. Dobzhansky and Spassky (1967), applying the same selection regime to *D. pseudoobscura*, could not obtain populations able to survive concentrations of salt greater than 3%.

Factors other than pleiotropy, particularly linkage and fixation of alleles, contribute to the plateauing in the response to directional selection. But pleiotropy may be the most likely explanation of a second common phenomenon. If directional selection is relaxed, the character selected may return to its original state. Figure 2 shows an example of both the plateauing and the subsequent effect of relaxing selection. The return to the original state can be explained by the effect of individual selection on other characters affected by the pleiotropic genes whose frequencies were altered by the directional selection. When the directional selection is relaxed, the individual selection would tend to return the pleiotropic genes to their original frequencies.

24

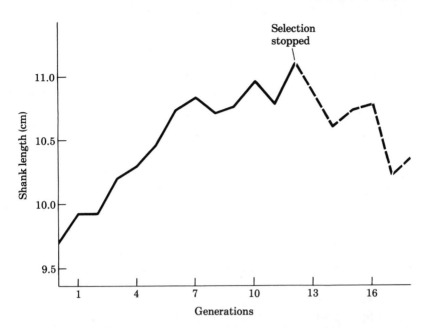

FIGURE 2. The results of directional selection on shank length in chickens for 12 generations with no selection after generation 12. Note that a rough "plateau" is reached by generation 7 and that after selection is relaxed the population average partly returns to the original value. (After Dempster, 1958, Figure 4.10.)

Pleiotropy may be due to a gene's effect on the timing of developmental events. Gould (1977) and Alberch et al. (1979) have emphasized the importance of this kind of pleiotropy both for producing large changes in several characters through genetic changes at a single locus, and for constraining genetic changes at loci that control the timing of development. The potentially large phenotypic effect of such genes may give them an especially important evolutionary role.

Linkage

In general, genes affecting the same character are not closely linked; and, conversely, closely linked genes affect different characters. Linkage between genes affecting different characters can cause selection on one character to cause a correlated change in other characters. For example, selection on color could result in a change in body size. Deleterious changes in the correlated characters could stop the re-

25

sponse to selection acting on the first character. Linkage among loci is thought to be another cause of plateauing in selection experiments (Dickerson, 1955).

Although linkage can be shown to be a constraining force in selection experiments, it is not clear how important linkage is in natural populations. Theoretical studies show that unless correlations among genes are actively maintained by selection, they will quickly decay, and the conditions under which permanent correlations are maintained in populations are restrictive (Ewens, 1979). Furthermore, there is little evidence that there are permanent correlations among genes in natural populations, except for genes in inversions (Charlesworth and Charlesworth, 1973; Langley, Smith and Johnson, 1978). Linkage is probably only a temporary constraint on genetic evolution both because correlations caused by linkage will decay on their own and because they can be broken up easily by selection.

Functional constraints

Two or more phenotypic characters may be constrained in their evolution because they are functionally dependent on one another. Adaptations to particular ecological conditions might well be incompatible. For example, ecological conditions experienced by seed-eating mammals may favor individuals able to eat larger and tougher seeds. Natural selection could then favor larger molars with more durable grinding surfaces. But for tooth size to increase significantly, the entire jaw structure must change. Selection on tooth size could be opposed by selection preventing changes in other parts of the jaw.

Some functional constraints are imposed by the laws of physics. Many of the well-known examples of allometry (Huxley, 1932) are a consequence of functional constraints imposed by the relationship between surface area and volume. For example, as an animal becomes larger and increases its length by some factor c, the cross-sectional area of its supporting limbs must increase by a factor of $c^{3/2}$ (Maynard Smith, 1968), unless the strength or number of limbs is increased accordingly. Similar considerations apply to the sizes of other organs such as lungs, heart muscle, and brain (Gould, 1966). In fact, J. B. S. Haldane (1928) suggested that much of functional morphology is the story of the struggle to increase surface area in proportion to volume.

Other functional constraints are due to the use of a character for more than one purpose. For example, the males of many bird species are brightly colored. Bright colors probably do not promote the survival of males, because females and juveniles are not also brightly colored. Yet the bright colors are maintained presumably by female preference for brightly colored males. Whatever advantage there is to duller colors does not change the coloration of the males because of the other function of the bright colors. In a similar way, the talons of birds

26

of prey might be improved as instruments to hold their prey if the talons did not also serve for walking and for grasping perches.

Functional constraints do not entirely prevent evolution of a character for a particular purpose, but they probably slow down the process of adaptation and limit evolutionary changes in any one direction. Allometric relationships can be modified by artificial selection (Atchley and Rutledge, 1980). Several comparative studies have shown that allometric relationships are sensitive to ecological factors (e.g., Baron and Jolicoeur, 1980; Harvey et al., 1980).

Adaptive valleys

The evolution of two or more characters could be constrained by the fact that both would have to be changed significantly before any advantage were gained. In terms of the adaptive surface, this situation is represented by two or more adaptive peaks separated by adaptive valleys (Figure 1). Although one adaptive peak might be higher than another, individual selection alone could not move a population to another peak. For example, if a particular plant species were defended against insect herbivores by two toxic compounds, a species attempting to exploit that plant would have to evolve ways of dealing with both. Even though the insect could develop ways of dealing with both toxins if presented with them one at a time (thereby allowing it to use a new resource), it might never do so because it could never succeed in dealing with both toxins together.

Wright (1931) argued that many major evolutionary transitions were of this type and that genetic drift, group selection, and other stochastic forces were necessary to explain the shift from one adaptive peak to another. This is Wright's "shifting balance" theory of evolution (Wright, 1955). The importance of adaptive valleys in constraining evolution and the importance of genetic mechanisms for crossing adaptive valleys are not agreed upon. Fisher (1930) followed Darwin in saying that the dominant mode of evolution was the continued improvement of characters and combinations of characters. Evolution proceeded up adaptive ridges. Darwin's discussion of the evolution of the vertebrate eye (1859, pp. 187–190) is still a good example of this view. It is often difficult to resolve the difference between this view and Wright's, because of the absence in the fossil record of intermediate forms whose fitnesses can be assessed. The dispute is often over the existence of plausible intermediates; and the failure to envision a sequence of functioning intermediate forms, as described by Darwin for the vertebrate eye, could be due to lack of imagination or to their nonfeasibility.

Changing environmental conditions

A character may fail to evolve because there has not been sufficient time for it to do so. The pattern of environmental variation a species experiences constrains its evolution by allowing adaptation only to those conditions that persist for some time. It is possible that many characters are adapted to past rather than present conditions. Gould and Lewontin (1979) have emphasized that if many characters are in fact not adapted to present conditions, attempts to prove that they are will lead to fallacious and incorrect conclusions. The composition of most communities can change rapidly through large fluctuations in population sizes and through colonizations and extinctions. The potentially short time scale of ecological change may not allow time for the coevolutionary response of most species.

Gene flow

If a species is widely dispersed and experiences different environmental conditions in different parts of its range, then gene flow can oppose the adaptation to local conditions. Sufficiently strong gene flow can completely swamp local adaptations and prevent the genetic divergence of local populations. Theoretical and experimental studies (reviewed by Endler, 1977) show that the swamping of local adaptations is indeed possible, but there is disagreement about how frequently it occurs in natural populations.

Mayr (1963) and more recently Stanley (1979) have argued that gene flow is generally important in preventing local adaptations and thereby causes an entire species to evolve as a single unit. Examples regarded as typical are populations of small mammals that have a dark pelage when living on lava flows in isolated populations but not when the populations on lava flows receive immigrants from populations living on lighter colored rocky substrates (Dice and Blossom, 1937). The view of Mayr and Stanley is that gene flow is generally a strong constraining force to selection in each population.

Ehrlich and Raven (1969) and Endler (1977) argue for the opposite view, namely that gene flow is usually a weak force relative to selection. There are two parts to their argument: first, that measured rates of gene flow are too low to swamp local adaptations; and, second, that isolated populations do not show substantially greater degrees of genetic divergence than do populations receiving immigrants. Ehrlich et al. (1975) describe their study of the butterfly *Euphydryas editha*, which is found both in western California and Colorado but not in the deserts in between. There has been no more phenotypic divergence of populations that have been geographically isolated for tens of thousands of years than there has been of populations known to exchange migrants.

Which view of the role of gene flow is adopted in a coevolutionary study determines the geographical scale that must be examined. If gene flow is relatively weak, then it is appropriate to examine interactions within each community. But if gene flow is relatively strong, then interactions over entire species' ranges would have to be considered.

RATES OF EVOLUTION

Simpson (1953) distinguishes between morphological rates of evolution (as measured by the rate of change in the measure of character) and taxonomic rates (as measured by the rates of turnover of taxa). For coevolution, morphological rates are very important, but they cannot usually be measured directly; therefore, morphological rates are often inferred from estimates of taxonomic rates.

Rates of morphological evolution can be measured precisely in artificial selection experiments. The conclusion from many experiments on many species is that most characters can evolve very quickly and that enough genetic changes occur to produce in a few generations a range of phenotypes that were not found in the initial population. In fact, it is unusual to find a character [such as development time in *Drosophila melanogaster* (Falconer, 1960)] that does not respond to selection for both increased and decreased values. Strong directional selection on a character usually produces correlated changes in several other characters.

Rapid evolution due to strong selection has also been observed in natural populations. The evolution of melanism, insecticide and disease resistance, and various weedy traits of plants have all been caused by human activities. Rapid morphological evolution has been inferred from high taxonomic rates in some groups. At least five species of moths in the genus *Hedylepta* have evolved to specialize on bananas in Hawaii, even though that plant was introduced only 1000 years ago (Zimmerman, 1960). Cichlid fishes have undergone extensive adaptive radiations during the past one or two million years in several African lakes (Stanley, 1979), and they have undergone extensive morphological changes during these radiations.

Rapid evolution, as measured by morphological or taxonomic rates, can occur under natural conditions, but it does not have to occur and certainly has not occurred in all lineages. The so-called "living fossils" such as the horseshoe crab or the alligator are well known but sometimes are regarded as relict species and not typical of species in more successful groups. A different view is expressed by the proponents of the punctuated equilibrium theory of evolution (Eldredge and Gould, 1972; Stanley, 1979). As Stanley (1979) extensively documents, the fossil record of several groups shows that many species do not undergo significant morphological changes once they appear. This observation has led to the theory that the typical pattern of evolution is the morphological stasis of widespread species, with rapid morphological evolution associated with species formation. As discussed by Stanley (1982) and Charlesworth et al. (1982), this pattern of the fossil record is not necessarily incompatible with the view that natural selection is the predominant mechanism causing both the stasis and the rapid changes during species formation, but Gould (1980) and Alberch (1980) have argued for the importance of other mechanisms, including developmental constraints and genetic drift.

If interactions among species are indeed strong, then rapid coevolution could result. The evidence that rapid morphological evolution can take place in nature shows that rapid coevolution is a possibility. But rapid coevolution should not be assumed without considering alternatives. Interactions may not be as strong as supposed and various constraints may prevent significant coevolutionary changes.

OPTIMIZATION AND GAME THEORY

It is rare to study a phenotypic character for which the genetic basis and the ecological role are both well understood. The evident reason for the use of the same few examples of genetic evolution—industrial melanism in *Biston betularia* or banding patterns in *Cepaea nemoralis*—is the scarcity of such examples. Most workers are faced with understanding the evolution of characters whose genetic basis is completely unknown and species for which breeding studies are impossible. This situation is common in the study of coevolution, and it is usual to follow the tradition of Darwin and construct coevolutionary hypotheses without explicit genetic information about the characters of interest. This approach is based implicitly on evidence that there is substantial genetic variation in characters whose genetic bases can be understood.

In the past two decades, several formal theoretical methods have been used to make inferences about phenotypic evolution in the absence of genetic information. Levins (1968) developed a theory of "fitness sets" similar to utility theory in microeconomics (Peterson, 1974) to predict the outcome of evolution in spatially or temporally varying environments. MacArthur and Pianka (1966) developed a theory of feeding behavior based on the principle that animals will choose an optimal diet. That theory has grown into the theory of "optimal foraging strategies" (Schoener, 1971b; Pyke, Pulliam, and Charnov, 1977). Hamilton (1966) implicitly used a game theory approach in defining an "unbeatable strategy" in his analysis of the evolution of senescence; and Maynard Smith and Price (1973) made direct use of

30

game theory to analyze the problem of fighting and displaying in social animals. Maynard Smith and Price (1973) introduced the term *evolutionarily stable strategy* (ESS) to describe the mixture of phenotypes in a population that would resist invasion by any other phenotypes.

These theoretical approaches have had great influence on both evolutionary biology and ecology, and for good reason. They show the common elements of seemingly unrelated systems and provide simple ways to characterize evolutionary trends without requiring detailed knowledge about the traits of interest. Both optimization theory and game theory can be useful for understanding coevolution. For example, optimal foraging theory can lead to predictions about the interactions between a predator and its prey or a parasite and its host. Game theory can be used to understand models of coevolving systems by envisioning each species as a "player" that is trying to maximize its gain through certain kinds of interactions under the assumption that other species are trying to do the same (Slatkin and Maynard Smith, 1979). For example, the coevolution of two competing species can be understood in terms of the strategy of generalization leading to the use of more resources but causing more intense competition, as contrasted with a strategy of specialization leading to the use of fewer resources but also leading to less intense competition (Lawlor and Maynard Smith, 1976).

Optimization theory and game theory also have their limitations, and recently their application to evolutionary problems has been criticized. One criticism is that the characterizations of phenotypic evolution obtained using one of these theories may be incorrect in the sense that there may be no genetic model that is consistent with the characterization. For example, Strobeck (1975) has shown that Levins' theory of fitness sets cannot predict the outcome of temporally varying selection of certain kinds.

Another criticism is that these theories may be correct only if additional assumptions are made. For the application of game theory, Slatkin (1979) and Maynard Smith (1981) have shown that it is necessary to assume that there is sufficient genetic flexibility of the characters of interest before game theory predicts the correct equilibrium states. The application of optimization theory to the evolution of some character requires the assumptions that the character of interest is under genetic control and that there are no constraints of the kind discussed earlier in this chapter that would prevent the optimal state from being reached. The difficulty is not that these assumptions are necessarily invalid but that the theories themselves have no way to test the validity of their underlying assumptions in

31

particular cases. These methods generally lead to some prediction about evolution and coevolution even when they are inapplicable.

An important problem in the application of optimization or game theory to evolution is that the range of possible phenotypes must be specified. The optimal foraging strategy of a predator depends on its speed of movement, its agility, its ability to discriminate prey, and its memory. With no limits on these factors, optimization theory would predict that a predator would move infinitely quickly to any point, possess perfect knowledge of every prey item, and remember all past prey. The actual predictions about the predator depend on what limits are assumed. Levins (1968) called the range of phenotypes the "fitness set" of the species. In game theory, the fitness set is the set of strategies that each player has to choose from. Often, the most difficult part of an evolutionary or coevolutionary model is the specification of the fitness set.

Some criticisms of the use of optimization theory and game theory in evolution are valid. As Gould and Lewontin (1979) point out, ecologists are reluctant to conclude that a particular trait is not optimally adapted to some purpose. Because the prediction of optimization and game theory can depend strongly on the fitness set assumed, it is usually not possible to show that they are inapplicable to a particular trait. Although they are not a substitute for detailed genetic and ecological information, these approaches can provide a useful way to summarize that information.

CONCLUSION

The neo-Darwinian theory of evolution generally provides support for the assumption implicit in most coevolutionary studies that any particular character can evolve quickly if the individual selection affecting that character is strong enough. The neo-Darwinian theory, however, provides additional information that suggests that some caution should be used and some alternatives considered. Genetic changes can result from selection at levels above the individual. Kin and group selection could be effective in a species that is divided into transient local populations. In such populations, individual selection may well not have time to act.

There are also many constraints on the evolution of particular characters. Constraints may be especially important for coevolution because of the specificity of the interactions assumed. The demonstration that species interact in a particular way does not itself imply that coevolution has occurred or will occur. It is possible that studies of coevolution will contribute to our understanding of the importance of selection at different levels and of the roles of genetic constraints in shaping evolutionary changes.

CHAPTER 3

THE THEORY OF COEVOLUTION

Jonathan Roughgarden

Coevolutionary theory is one of the newest areas of theoretical ecology. During the last ten years it has provided interesting, and sometimes important, insights into how coevolution works. Yet many questions remain to be explored. This chapter is an invitation to think about coevolution in theoretical terms. It offers a guide to the approaches that have been developed so far. More technical detail may be found in reviews by S. A. Levin (1978) and Slatkin and Maynard Smith (1979) and in books by Christiansen and Fenchel (1977) and Roughgarden (1979).

Coevolutionary theory lies at the interface between population genetics and theoretical ecology. Indeed, models of coevolutionary phenomena tend to highlight differences between the points of view held by scientists of these fields. Geneticists are primarily concerned with the influence of the mechanism of inheritance on the coevolutionary process and with the influence of population interactions on the genetic structure of a population. Ecologists are primarily concerned with the effect of coevolution on the phenotype, especially on traits that determine how populations interact, and with the effect of coevolution on the abundance and distribution of the interacting populations. My orientation is primarily ecological, but I have included references to more genetical approaches.

There are four main parts to this chapter. The first presents some thoughts about the purpose of theory and introduces some basic technical concepts about evolution. The second is an introduction to density-dependent evolution for a single population. The third is a survey of special cases of the coevolution between two species. The

33

fourth is a discussion of how coevolution may influence an entire biological community.

PART 1. PRELIMINARY THOUGHTS

Characteristics and uses of theory

There are two types of mathematical models in ecology. One type presents a simplified and possibly idealized picture of how processes operate. Its purpose is to develop insight. A simplifying model is an analogy rather than an approximation to what is actually going on.

Another type presents a summary of what is known. A summarizing model is useful when numerical predictions are needed, and when large amounts of data need to be rendered in a compact form.

The two types of models need not be mutually exclusive but usually are. Typically, the simplifications in a simplifying model make them too inaccurate for numerical applications, and a summarizing model is too complicated to explore mathematically.

There are no laws about coevolution in the sense of Newton's laws. Theory in ecology does not spring from axioms; instead, it is a small, but growing, collection of models that have proved useful to people.

Most people readily appreciate the value of a summarizing model because there is obvious utility in representing a body of data in a concise and transportable form. Volumes of data on tidal heights at the seashore and on solar illumination during the day are summarized with computer programs that are much easier to use than the original data tables. In contrast, simplifying models often produce a sense of uneasiness; if the assumptions are unrealistic, how can a model be of any value?

But experience shows that simplifying models really are helpful. From models of simple machines like the frictionless pulley and the inclined plane we learn the concept of mechanical advantage. These models are the starting point for the design of real machines. Similarly, in electronics, the formulae for the properties of resistors, inductors, and capacitors are as idealized as the model for a frictionless pulley and yet are used as building blocks in designing electrical circuits. In practice, simplified and idealized models are useful guides to understanding processes even though they use unrealistic and obviously false assumptions like that of no friction.

A conclusion from a model is robust if later research shows that it was originally derived from premises that were unnecessarily restrictive. In fact, almost all theoretical results are first derived in a restricted context that is later generalized to some extent. If the assumptions seem very restrictive, it is possible that the conclusions have limited interest, but it is also possible that the conclusions are more robust than the original assumptions suggest.

Absolute fitness

The classic models for evolution are based on an extremely simplified picture of what a population is and how it evolves (see Roughgarden, 1979, pp. 26–30). The life history is assumed to consist of discrete generations, with mating occurring through the random union of gametes. A population with this life history is assumed to live more or less in one place, so that the complexities of migration do not have to be considered. For this model population, let m be the fertility of an adult individual (m is one-half of the gametes that eventually become incorporated into zygotes). Let l be the probability that a zygote survives to be an adult. Then the *absolute fitness* for an individual is the product of the fertility with the survival probability.

$$W = ml \tag{1}$$

The units of W are numbers of individuals. W is zero or positive and is the expected number of offspring that the individual will contribute to the next generation.

Natural selection means that there are *differences* in absolute fitness among the individuals in a population. In population genetics, the *relative fitness*,

$$w = \frac{ml}{W_{\max}} \tag{2}$$

is usually used. Here $W_{\max}$ is the highest absolute fitness that any individual has. The *relative fitness* of an individual, w, must lie between 0 and 1 and is a unitless number because it is a ratio in which the units of absolute fitness cancel out. Traditional population genetics is based on analyzing relative fitnesses.

For our purposes, the approach of analyzing relative fitnesses is obsolete, because the absolute fitness contains the information needed to predict changes in the population size, and this information is lost when dealing with relative fitnesses. Let $\overline{W}_t$ denote the average absolute fitness in the population at time t. Because $\overline{W}_t$ is the average number of offspring produced per individual, the population size at time $t + 1$ (where time is measured in number of generations) is simply

$$N_{t+1} = \overline{W}_t N_t \tag{3}$$

The average absolute fitness $\overline{W}_t$ depends on the composition of the population—because it is an average over the different kinds of individuals in the population—and also on the population size itself. In a limited environment, the number of offspring left by an individual depends somehow on the resources available to that individual. If the average fitness at time t happens to equal 1, as it will if the population is near a size commensurate with its resources, then the population size at $t + 1$ will equal that at time t.

The fundamental theorem of natural selection

A famous result from traditional population genetics theory is that natural selection brings about the maximization of the average *relative* fitness in the population, that is, the relative fitness as defined in equation (2), averaged over all the individuals in the population. According to this result, the average relative fitness increases after each generation—rapidly if there is a great deal of difference among the individuals in their relative fitness and slowly if most of the individuals have nearly the same relative fitness. Regardless of speed, the average relative fitness continually increases as it approaches, though never attains, a peak value. This finding is called the "Fundamental Theorem of Natural Selection" (see Chapter 2 by Slatkin).

The fundamental theorem is very important, yet it is delicate and easily misunderstood. Evolution involves much more than natural selection; other components to the evolutionary process include mutation, recombination, segregation distortion, genetic drift, mating behavior, age structure, population distribution, correlations between different characters, mechanical constraints on morphology, and so forth. Thus, the fundamental theorem of natural selection is not the fundamental theorem of evolution. Nevertheless, it is approximately correct when selection is much stronger than these other factors.

However, the kind of natural selection to which the fundamental theorem refers is very limited; it pertains to a population in which the *relative fitness* of each individual is *constant through time*. It is not appropriate if the relative fitnesses depend on the population composition (frequency-dependent selection) or size (density-dependent selection). Yet these are precisely the forms of selection that naturally arise in the context of coevolution. There are, instead, replacements for the fundamental theorem that are useful tools in visualizing how frequency-dependent and density-dependent selection operate.

Evolutionarily stable strategy

One approach for frequency-dependent selection is to consider a population all of whose members have a given phenotype and then to ask whether there is a different phenotype that can enter into the population. If there is no other phenotype that can enter the population, then the given phenotype is said to be an *evolutionarily stable strategy*, or ESS (Maynard Smith and Price, 1973). When this approach is used, the set of all possible phenotypes must be precisely defined. The theoretical task is to calculate which phenotype, among all those possible, satisfies the condition that a population of such a phenotype cannot be invaded by individuals with any other phenotype. To do this, one finds the phenotype that has the highest fitness, given that it is the only type in the population. There may be no solution, one solution, or several solutions to this problem, depending on the detailed situation.

The justification for this approach is that fitness relations do predict the initial increase of a gene, even with frequency-dependent selection. The average fitness may decline as the gene continues to spread, but whether the gene enters in the first place depends on its carrier having a higher fitness than the other members of the population at that time (see Roughgarden, 1979, pp. 53–54). This justification also brings out the limitation of the ESS concept. It only refers to the initial increase of a phenotype when it is rare. To determine its subsequent fate, we need more information.

In contrast to the ESS approach that is focused on whether a monomorphic population is stable, Slatkin (1979) has concentrated on a polymorphic population. He has pointed out that a polymorphic population with frequency-dependent selection is, in some models, a state for which the fitness of every phenotype is equal and that there is a tendency toward the equalization of fitnesses.

PART 2.
DENSITY-DEPENDENT EVOLUTION

What is the connection between evolution and abundance? The main result from the theory of density-dependent evolution is that the evolution of traits that bring an increase in fitness causes an increase in abundance. This result seems intuitive, yet it is not generally true for coevolving populations even though it is true for density-dependent evolution in a single population.

Some traits of an organism promote a high growth rate for the population under conditions of low density (high-r traits). Other traits promote a high equilibrium population size for a population under conditions of high density (high-K traits). Density-dependent selection causes the evolution of high-K traits, and density-independent selection, which occurs under low density when a population is expanding, causes the evolution of high-r traits. These results can be derived using an extension of the idea of an evolutionarily stable strategy to density-dependent selection.

Evolutionarily stable ecological equilibrium

When the selection depends on population size, there are two kinds of variables in the system. The first (X) is a variable referring to the trait undergoing evolution. The second (N) is the population size itself, for this inherently changes as the traits whose fitness depend on population size evolve. The absolute fitness for an individual is a function of both variables, that is,

$$W = W(X,N) \tag{4}$$

The idea of an evolutionarily stable strategy with density-dependent selection is that $W(X,N)$ is maximized *with respect to X only,* and the magnitude of W is set equal to 1, as is appropriate if the population size is approximately constant. If a population consists of individuals whose trait maximizes fitness in this sense and if the population size is holding approximately constant through time, then any different type of individual introduced into this population cannot increase when rare. Such a population is at an *evolutionarily stable ecological equilibrium.*

To find the value of a trait that produces an evolutionarily stable ecological equilibrium, write down the fitness function $W(X,N)$ explicitly. Then take the partial derivative of $W(X,N)$ with respect to X and set the result equal to 0. This produces an equation in both X and N. Next, set $W(X,N)$ equal to 1. This is a second equation in X and N. Find the simultaneous solution to both equations. The solution requires that N be greater than 0 and that the second partial derivative of $W(X,N)$ with respect to X be negative. If N is not positive, then the solution is meaningless. If the second derivative is negative, then the solution is a maximum.

For example, suppose the absolute fitness of an individual decreases as a linear function of population size. This assumption leads to the logistic equation, which in discrete time may be written as

$$N_{t+1} = [1 + r - (r/K)N_t]N_t \tag{5}$$

where r is the intrinsic rate of increase and K is the carrying capacity. The expression in brackets is the absolute fitness W. Now suppose, further, that there is a trait whose value, X, influences an individual's r and K, so that we can write $r = r(X)$ and $K = K(X)$. The fitness then becomes

$$W(X,N) = 1 + r(X) - [r(X)/K(X)]N \tag{6}$$

For example, the trait X may refer to body size, and body size may be an indication of the average prey size that an individual uses. If so, it is convenient to take X as the logarithm of body size, and then $K(X)$ must be a function that tends to 0 as X tends to $+\infty$ and to $-\infty$. For

38

simplicity, we may view $K(X)$ as a unimodal function more or less like the familiar Gaussian bell-shaped curve. The peak of $K(X)$ roughly corresponds to the body size that yields the greatest net rate of prey capture. Similarly $r(X)$ depends on body size. What body size should evolve under this density-dependent selection?

The answer is found as follows: First we differentiate $W(X,N)$ with respect to X and set the result equal to 0.

$$\frac{\partial W(X,N)}{\partial X} = \frac{dr(X)}{dX} - N \left\{ \frac{K(X)\,[dr(x)/dX] - r(X)\,[dK(X)/dX]}{K(X)^2} \right\} = 0 \tag{7}$$

Next we set $W(X,N)$ equal to 1,

$$W(X,N) = 1 + r(X) - [r(X)/K(X)]N = 1 \tag{8}$$

Now, equation (8) simplifies to

$$N = K(X) \tag{9}$$

and then equation (7) reduces to

$$dK(X)/dX = 0 \tag{10}$$

Furthermore, taking second derivatives shows that $\partial^2 W/\partial X^2$ is negative if $d^2K(X)/dX^2$ is negative. So, the body size that evolves is the body size that maximizes the carrying capacity. Notice that the relation between X and r is unimportant; it is the relation between X and K that is critical. This point is the essence of what is called "K-selection." Under density-dependent selection what is important is only the equilibrium population size attained and not the rate at which that equilibrium population size is approached.

Evolutionary dynamics

When a population enters a habitat it presumably does not already possess the traits that produce an evolutionarily stable ecological equilibrium there. What happens between the time it enters and the time it achieves an evolutionarily stable configuration? The answer depends on many factors, including the actual genetic mechanism underlying the traits. There is a simple model for the evolutionary dynamics that seems to be useful, although it is based on ignoring the genetic and demographic complexities that are undoubtedly present. There are three main assumptions in the model: (1) evolutionary changes in the traits occur much more slowly than changes in population size; (2) the evolution of the traits can be described by formulae that originate in the study of plant and animal breeding where traits, like body size and shape, are caused to evolve through artificial selection; and (3) the population variance for the traits changes slowly relative to changes in the mean.

In the genetics of plant and animal breeding, the change in the mean value of a trait after one generation of selection is written as

$$\Delta \overline{X} = h^2(\overline{X}_{w,t} - \overline{X}_t) \tag{11}$$

where $\overline{X}_{w,t}$ is the mean among the adults after the selection, $\overline{X}_t$ was the mean before selection, $\Delta \overline{X}$ is the difference between $\overline{X}_{t+1}$ and $\overline{X}_t$, and h^2 is a constant between 0 and 1 that is called the heritability (see Falconer, 1960; Roughgarden, 1979, Ch. 9). Let the selective value of an organism of size X be $W(X)$. If $W(X)$ is expanded to first order about $\overline{X}_t$, we obtain

$$\overline{X}_{w,t} = \overline{X}_t + \frac{\sigma^2}{W(\overline{X}_t)} \frac{dW(\overline{X}_t)}{dX} \tag{12}$$

where σ^2 is the population variance in the trait. σ^2 is taken as a constant because it tends to a limit as long as the mean remains within the bounds of the variation that was originally present (see Roughgarden, 1979, pp. 142–145; Karlin, 1979). Putting these together, we obtain

$$\Delta \overline{X} = \frac{h^2 \sigma^2}{W(\overline{X}_t)} \frac{dW(\overline{X}_t)}{dX} \tag{13}$$

Suppose that the population can approach an equilibrium size, corresponding to a population in which the mean value of the trait equals X. This is found by setting $W(X,N) = 1$ and solving for N as a function of X. Then the model is

$$\Delta X = (\text{const}) \left. \frac{\partial W(X,N)}{\partial X} \right|_{N = \hat{N}(X)} \tag{14}$$

where const is a positive constant. In this model the population size "tracks" the current state of the evolving trait. By this model the trait evolves from any initial condition by climbing the gradient in the fitness function and comes to equilibrium at a value that represents an evolutionarily stable ecological equilibrium according to the previous definition.

This approach assumes that the population dynamics lead to a stable equilibrium abundance for any particular value of the trait. When this assumption is not reasonable, a specialized model for the dynamics must be used in order to characterize the nonequilibrium trajectories.

Other models for the evolutionary dynamics may be based on assuming a particular genetic mechanism for the trait, such as a single locus with two alleles (Anderson, 1971; Roughgarden, 1971; Clarke, 1972). In this particular case, a stable population genetic equilibrium brings about a local maximum to the equilibrium population size, as MacArthur (1962) had originally conjectured. Relevant literature now includes Roughgarden (1976), León and Charlesworth (1976), and Ginzburg (1977). References on r- and K-selection include Cody (1971), Gadgil and Solbrig (1972), and Harper (1977).

PART 3. COEVOLUTION BETWEEN TWO POPULATIONS

Coevolutionarily stable community

Coevolution is the simultaneous evolution of ecologically interacting populations. These interactions include interspecific competition; plant–herbivore and predator–prey interactions; and symbiotic relationships of parasitism, commensalism, and mutualism. For coevolutionary ideas to apply in a particular system, the populations in that system must interact, or have interacted in the past, and must have been together long enough in space and time for the interaction to have had a realistic opportunity to cause evolutionary changes.

We can extend the ideas first encountered in density-dependent selection to the coevolution of interacting populations. Consider two species, with population sizes N_1 and N_2, respectively. Suppose also that there is a trait carried by members of Species 1 whose state is X_1, and also a trait in Species 2 whose state is X_2. Then the fitness of an individual in each species depends on all of these variables. Natural selection within Species 1 responds to the traits of other species but modifies only the traits within Species 1 itself. Hence, suppose we

$$\text{maximize } W_1(X_1,X_2,N_1,N_2) \quad \text{with respect to } X_1 \text{ and set } W_1 = 1$$
$$\tag{15}$$
$$\text{maximize } W_2(X_1,X_2,N_1,N_2) \quad \text{with respect to } X_2 \text{ and set } W_2 = 1$$

A solution to this problem yields a value for the trait in Species 1 and a value for the trait in Species 2 that maximize the fitness within each of those species, respectively, and also yields the population sizes for both interacting species commensurate with their joint use of the resources in the environment.*

A community that is determined in this way is a *coevolutionarily stable community* (CSC). It is stable both ecologically and evolutionarily in the following sense: If every member of Species 1 has a trait whose value is X_1 and if every member of Species 2 has a trait whose value is X_2, and if the population sizes are N_1 and N_2, where

---

*The ecological feasibility of the solution should also be checked and the matrix $\partial W_i/\partial N_j$ should have eigenvalues consistent with ecological local stability.

(X_1,X_2,N_1,N_2) were obtained from the simultaneous solution of the equations in (15), then a mutant with some other value of X_1 cannot increase when rare in Species 1, and also a mutant with some other value of X_2 cannot increase when rare in Species 2.

To obtain a phenotypic model for the coevolutionary dynamics leading to a CSC, we can suppose that the population sizes "track" the current values of the traits in the populations, yielding

$$\Delta X_1 = (\text{const}_1) \left. \frac{\partial W_1(X_1,X_2,N_1,N_2)}{\partial X_1} \right|_{\substack{N_1 = \hat{N}_1(X_1,X_2) \\ N_2 = \hat{N}_2(X_1,X_2)}}$$

$$\Delta X_2 = (\text{const}_1) \left. \frac{\partial W_2(X_1,X_2,N_1,N_2)}{\partial X_2} \right|_{\substack{N_1 = \hat{N}_1(X_1,X_2) \\ N_2 = \hat{N}_2(X_1,X_2)}} \tag{16}$$

This system is used in Roughgarden et al. (1982) to model the dynamics of the coevolution of competitors. The results relate to the coevolution of the *Anolis* lizard populations on islands in the eastern Caribbean as discussed in Chapter 17.

An important limitation of this approach is the assumption of a stable population dynamic equilibrium for every combination of traits in the species; the approach is restricted to the evolution of model parameters where this assumption is true. In multispecies models, there are increasingly frequent findings of parameter sets that lead, asymptotically, to complicated nonequilibrium trajectories (e.g., May and Leonard, 1975; May and Anderson, 1978; and Arneodo et al., 1982).

The coevolution of competitors

How does coevolution influence two competing species? The issue is whether competition causes the evolution of resource partitioning, as indicated, for example, by a difference in the body sizes of the organisms. Biological examples of this idea are considered in more detail in Chapter 17. Here we shall treat this problem theoretically.

Suppose, as before, that the body size of Species 1 is X_1 and of Species 2, X_2. Let $K(X)$ be the carrying capacity of a species as a function of its body size. The new assumption is that the competition coefficient between individuals of different species is assumed to be a function of the difference in their body sizes. Specifically, the competition coefficient for the effect of an individual of Species j against an individual of Species i is

$$\alpha_{ij} = \alpha(X_i - X_j) \tag{17}$$

where $\alpha(X_i - X_j)$ is called the competition function.

To illustrate, we assume the competition function is a Gaussian bell-shaped curve whose peak occurs where $X_i = X_j$. When $X_i = X_j$, we

42

let $\alpha = 1$, indicating that inter- and intraspecific competition are equal. The width of the competition curve is described by the parameter σ_α. The competition function is

$$\alpha(X_i - X_j) = \exp[-(1/2)(X_i - X_j)^2/\sigma_\alpha^2] \tag{18}$$

Similarly, we let the carrying-capacity curve be a Gaussian, with X scaled so that the peak lies at $X = 0$,

$$K(X) = K_m \exp[-(1/2)X^2/\sigma_K^2] \tag{19}$$

For simplicity, we take r for each species as a constant independent of X, but identical results arise if both r_1 and r_2 are functions of X_1 and X_2, respectively.

The fitness functions are

$$W_1(X_1,X_2,N_1,N_2) = 1 + r_1 - [r_1/K(X_1)]N_1 - \alpha(X_1 - X_2)[r_1/K(X_1)]N_2 \tag{20}$$

$$W_2(X_1,X_2,N_1,N_2) = 1 + r_2 - [r_2/K(X_2)]N_2 - \alpha(X_2 - X_1)[r_2/K(X_2)]N_1$$

To find the coevolutionarily stable community based on these fitness functions, we differentiate W_1 with respect to X_1 and set the result equal to 0. Then we set W_1 equal to 1, thereby providing two equations based on W_1. Similarly, we get two more equations from W_2, yielding four equations in four unknowns:

$$N_1 d[1/K(X_1)]/dX_1 + N_2 \partial[\alpha(X_1 - X_2)/K(X_1)]/\partial X_1 = 0$$

$$K(X_1) - N_1 - \alpha(X_1 - X_2)N_2 = 0$$

$$N_2 d[1/K(X_2)]/dX_2 + N_1 \partial[\alpha(X_2 - X_1)/K(X_2)]/\partial X_2 = 0 \tag{21}$$

$$K(X_2) - N_2 - \alpha(X_2 - X_1)N_1 = 0$$

Because of the symmetry in the $\alpha(X_i - X_j)$ and $K(X)$ functions, we know there is a solution of the form $N_1 = N_2 = N$, and $X_1 = -X_2 = X$. The top equation in (21) then becomes

$$N\left\{ d[1/K(X)]/dX + \partial[\alpha(X_1 - X_2)/K(X_1)]/\partial X_1 \Big|_{\substack{X_1 = X \\ X_2 = -X}} \right\} = 0 \tag{22}$$

The last term is written to indicate that the competition function is first differentiated with respect to X_1, and then X_1 and X_2 are set equal to X and $-X$, respectively. If N is positive, then the expression in braces must equal zero. Upon substituting the carrying-capacity and competition functions explicitly, and solving, we obtain

$$X/\sigma_\alpha = + [(1/2)\ln(2\sigma_k^2/\sigma_\alpha^2 - 1)]^{1/2} \tag{23}$$

43

provided $\sigma_k > \sigma_\alpha$. Meanwhile, the equilibrium population sizes are found to be

$$N = K(X)/[1 + \alpha(2X)] \tag{24}$$

Thus, the coevolutionarily stable community for these symmetrically competing species consists of one species whose body size is X and another species whose body size is $-X$, and whose population sizes are N where X and N are given by (23) and (24) above. The evolutionary dynamics of the approach to this equilibrium can be modeled with the equations for ΔX_1 and ΔX_2 given previously in (16).

This mathematical example illustrates an approach to determining how coevolution between competitors can cause the formation of resource partitioning. Resource partitioning between species may also be caused by the nonevolutionary mechanism of differential invasion. Hence, the existence of resource partitioning between competitors is not itself compelling evidence of coevolution. Also, the state of resource partitioning predicted in the mathematical example does not need to be attained through evolutionary divergence by the two species. Whether parallel or divergent evolution occurs depends on the initial condition for the coevolution. It is true that the end point can be viewed as a displacement by each species in an opposite direction away from the body size that either would evolve in the absence of the other. However, this view has nothing to say about the actual evolutionary trajectories taken by the coevolving species. These points are discussed further in Chapter 17.

Some of the recent theory on the coevolution of competitors examines the robustness of the result presented in the preceding mathematical example. Key references include Roughgarden (1972, 1976), Bulmer (1974), Fenchel and Christiansen (1977), Slatkin (1980), and Case (1982). A frequent theme concerns the *simultaneous* evolution of average differences between species together with differences among members within each of the species themselves. The discussion concerns traits, like body size, that might be taken as indicators of resource partitioning by individuals between and within species. These papers differ principally in the way that the inheritance of the character is treated. They offer several theoretical arguments that the evolution of resource partitioning between species requires a constraint on the evolution of the variance within each of the species. Furthermore, Roughgarden (1972) and Fenchel and Christiansen (1977) suggest that the evolution of an increase or decrease of the variance of a character within a species is a much slower process than the evolution of a shift in the mean value of the character. Roughgarden (1974), Hespenheide (1975), and Case (1982) argue that the empirical evidence shows the population variance to be evolutionarily conservative relative to the mean of the character. Case (1982) also offers a promising approach toward combining competition theory with the theory of optimal foraging.

If the competition function is asymmetric, then the outcome of coevolution is not necessarily the attainment of niche separation (Roughgarden et al., 1982). Instead, one of the competing species can be driven to extinction by the other species during the course of coevolution. If the community is reinvaded after this extinction, the process can repeat itself. The sequence of coevolutionarily driven extinction followed by invasion, if repeated, produces a long-term turnover in the residents of a community while preserving a total species diversity that is approximately constant. This cyclic process is discussed further below and in Chapter 17.

The extension of the mathematical example to multiple resource axes shows that alternative patterns of resource partitioning can be simultaneously coevolutionarily stable (Pacala and Roughgarden, 1982). This result means that a community can develop a pattern of resource partitioning on either one axis or another, and the result will be coevolutionarily stable in either case. Which pattern actually develops in a particular community depends on historical factors, including any predisposition to a specific axis brought by the original colonists from their ancestral habitat. This result may explain the phenomenon of "niche axis complementarity" discussed by Schoener (1974), whereby the identity of the axis used for resource partitioning varies from place to place. An example of niche complementarity in eastern Caribbean anoles is presented in Roughgarden et al. (1982).

Habitat segregation refers to the occurrence of species in different habitat types. What is being considered are large-scale habitats, not microhabitats. For example, in the Sierra Nevada, Clark's nutcracker occurs at mountain tops and is replaced by the Steller's jay at middle elevations, and by the scrub jay at low elevations. The area of the habitats involved is on the order of thousands to more than one million times the area of the home range of an individual.

If the environment has variation on such a large spatial scale to begin with, we may ask whether the population will tend to occur more in one part of its potential range than another. When genes enter the population that confer higher fitness in one part of the potential range at a cost in fitness in other parts of the range, the population abundance shifts to the favored region of the environment. Thus, even a solitary population in a spatially varying environment can evolve habitat segregation through the accumulation of genes that cause there to be more individuals in one part of the potential range than other parts. If this evolution occurs in a reciprocal manner by two com-

45

peting species, then the resulting pattern is a replacement of competitors along a large-scale gradient like the altitudinal gradient in the Sierra.

To model the evolution of habitat segregation on such a large spatial scale, we need to consider that local migration is possible only between nearby points within the potential range. A standard competition model may be used to provide a picture of the ecology and evolution at local points within the range. Next, the local models for nearby points are coupled by assuming that migration leads to a flow of individuals back and forth. The local population-interaction models have parameters that vary from place to place within the potential range of the populations.

A regional model of this form shows that the spread of a gene, one that increases the abundance of individuals at some points in the range while decreasing it at others, is governed by a threshold (Nagylaki, 1975; Fife and Peletier, 1981; Roughgarden et al., 1982). The threshold is calculated from the geometry of the habitats within the potential range, the degree of difference between the habitats themselves, and the average dispersal distances of individuals. The existence of a threshold means that the evolution of habitat segregation in some regions will not occur at all unless the region satisfies a certain condition.

This result may relate to why some regional systems develop habitat segregation and others do not. Cody and Mooney (1978) have shown that the birds of the chaparral in Chile do not have the degree of habitat segregation that exists among the chaparral birds of California. Also, the *Anolis* lizard populations from some islands in the eastern Caribbean have evolved habitat segregation whereas others have not (Roughgarden et al., 1982).

In summary, the theory for the coevolution of competitors offers results mostly about how competition can cause the evolution of species differences in traits that promote resource partitioning. Other results concern the simultaneous evolution of between- and within-species differences, the invasibility of a coevolved community, species extinction as a result of coevolution with asymmetrical competition, niche axis complementarity, and habitat segregation. There is not nearly as much theory for coevolution involving other ecological interactions. This is not a value judgment on the importance of competition relative to other population interactions; it is a historical accident. I now turn to one of the most important interactions—the predator–prey interaction.

Predator–prey coevolution

The predator–prey interaction is a conflict in which the predators kill prey for food, and yet are dependent upon the continued existence of the prey population for their own survival. Hence, we may ask the following questions. (1) Does evolution of more efficient predators through time eventually lead to predator extinction because of over-exploitation of the prey? (2) Is evolution of increasing predator ability counteracted by a continual evolution of defense, evasion, or toxicity by the prey so that a steady state is maintained? This is sometimes called the "Red Queen" hypothesis: both predator and prey are running hard just to stay even with one another. (3) Does the coevolution of predator and prey result in a coevolutionarily stable community, with an evolutionarily stable expenditure by predators to catch prey and by prey to avoid capture? Each of these questions leads to still more questions in the same vein. Although more work remains to be done, something can be said about these questions now.

The effect of predator–prey coevolution on the stability of the predator–prey interaction itself can be classified for each parameter in a predator–prey population dynamic model. Suppose, for example, that the prey grow logistically and that the predators have a linear functional response (an individual predator's predation rate in relation to the abundance of prey) with a numerical response (population growth rate of the predator in relation to prey abundance) proportional to their functional response (see, for example, Roughgarden, 1979, pp. 440–442). Then the fitnesses for the prey and predator are

$$W_v = 1 + r - (r/K)V - aP \tag{25}$$

$$W_p = 1 - \mu + abV \tag{26}$$

where V and P are prey and predator population sizes, respectively; r is the growth rate of the prey; μ is the density-independent death rate of the predator; a is the slope of the predator's functional response; and b relates the numerical response of a predator to its functional response. There is an equilibrium at which both predators and prey may coexist,

$$\hat{V} = \mu/ab, \quad \hat{P} = (r/a)[1 - \mu/(abK)] \tag{27}$$

This is globally stable for any positive initial condition provided P is positive. The $\hat{P}$ is positive and hence coexistence of predator and prey occurs if $abK > \mu$.

By inspection of W_v we see that natural selection within the prey will tend to increase r and K and to decrease a, everything else being equal. Notice, however, that the evolution of the prey's r and K will have no effect on prey abundance but will raise the predator population size. Evolution that lowers a will increase $\hat{V}$, lower $\hat{P}$ provided

47

$\hat{V} > K/2$ to begin with, and raise $\hat{P}$ if $\hat{V} < K/2$ to begin with. The highest value of $\hat{P}$ occurs at $a_0 = 2\mu/(bK)$, the value that produces $\hat{V} = K/2$.

Similarly, evolution in the predators will tend to lower μ and increase the product, ab, everything else being equal. The decrease in μ will lower $\hat{V}$ and raise $\hat{P}$, the increase in ab lowers $\hat{V}$ but may lower or raise $\hat{P}$.

With this model we can ask whether the predators will tend to evolve themselves to extinction. Clearly, a positive $\hat{P}$ requires that $abK > \mu$. Since the effect of evolution within the predator is to raise ab and to lower μ, the predator, if initially capable of coexisting with the prey, cannot in this model evolve so as to drive itself to extinction through overexploitation of the prey. (However, this is a deterministic result. If the predator evolves a large enough ab, the prey population size $\hat{V}$ may become very small, so that the prey could become extinct because of random factors. Extinction of the predator would then follow.)

The main destabilization to predator–prey coexistence comes from evolution in the prey. First, as the prey evolve defenses, a becomes lower, so the condition for positive $\hat{P}$, namely $abK > \mu$, may become violated. Hence, the prey will have evolutionarily evaded the predator. Second, as the prey evolve a higher K, the system tends to become more oscillatory. In more complex models (see, for example, Roughgarden, 1979, pp. 443), a stable limit cycle exists for K sufficiently high. In such a situation, the predator or prey population may become extinct, by chance, as population size comes close to 0 during the cycle. This point is the "paradox of enrichment" discussd by Rosenzweig (1971). The paradox is that an experimental augmentation of the prey's productivity may cause the predators to go extinct rather than benefit from the enhancement of the prey's productivity.

The idea that the evolution of increased predator ability is matched by the evolution of increased ability among the prey to defend against and to evade predation is treated in a series of papers beginning with Rosenzweig (1969, 1973) and culminating in Schaffer and Rosenzweig (1978). The argument is very complicated but can, I believe, be fairly restated as follows.

Consider the predation coefficient a in the predator–prey model. We may formally partition the differential of a, representing a small net evolutionary change per generation, da, into two components

$$da = \delta_v a + \delta_p a \qquad (28)$$

where $\delta_v a$ is the change to a caused by evolution in the prey and $\delta_p a$ is the change to a caused by evolution in the predators. As before, $\delta_v a < 0$ and $\delta_p a > 0$, everything else being equal. Schaffer and Rosenzweig argue that the overall da is positive if a is sufficiently small and

48

that in some cases da is negative if a has an intermediate positive value. If so, there is at least one value of a such that $-\delta_v a = \delta_p a$, representing a state wherein the continual evolution of increased predator abilities is being matched by counterevolution in the prey. This state results in no net change in the realized predation rate.

The reason for asserting that da is positive when a is low is that in this situation $\hat{V}$ is near K and so there are very few predators available. Hence, the selection pressure within the prey to avoid a rare predator is low, whereas the selection pressure within the predators to increase predation on an abundant prey population is high. Hence, the net effect should be to increase a, if a is sufficiently low.

On the other hand, if a is high, the argument becomes complicated. If a is high enough, there are few prey and also few predators, since for sufficiently high a, both $\hat{V}$ and $\hat{P}$ vary inversely with a. Hence, the net effect on δa is hard to gauge when a is very big. But if a has an intermediate value, one near that producing the highest predator population size, then the selection pressure within the prey to avoid predation will be very high. The value of a producing the highest predator population size is $a = 2\mu/(bK)$. In this case, it may turn out that the strength of this selection to avoid predation exceeds the strength of selection within the predators to increase predation on prey of intermediate abundance. If so, the net effect will be to lower a.

Thus, if a is near 0, da is positive. And if a is near the value producing the highest predator population size, then da may be negative. If so, the system may come to a dynamic coevolutionary steady state where the predation rate parameter a is between 0 and $2\mu/(bK)$ and where each generation the predators improve upon their ability to catch prey but are countered by improvements among the prey in their ability to avoid predation.

This treatment pertains to the accumulation of new or novel mutations. Schaffer and Rosenzweig suppose that in both species there is a finite mutation rate per generation that continually supplies these novel mutations. Since the strength of the selection in each species on these mutations varies with the ratio of $\hat{V}$ to K, it is possible, though not certain, that a steady state results. Obviously, the major unknown is whether mutation rates increasing a in the predators and decreasing a in the prey are high enough for the steady state to be an important coevolutionary phenomenon in nature.

The other approach to predator–prey coevolution deals with rearrangements of the basic phenotypic plan in both predator and prey. It is assumed that there are no novel mutations available to make predators and prey unconditionally better at predation and avoidance.

Instead, for a predator to catch more prey, it must pay a price in terms of other activities. Similarly, the prey must pay a price for the manufacture of defensive toxins and so forth. There is assumed to be in each species genetic variation that permits the rearrangement of the prey's phenotype and the predator's phenotype subject to constraints inherent in the basic body plan of each species. This approach yields a prediction of a coevolutionarily stable community (CSC) for predator and prey in which there is an evolutionarily stable level of effort expended by predators in their search for prey and by prey in their avoidance of predation.

Suppose, for example, that the prey have a trait X_v that can be interpreted as a measure of prey defense against predation and that the predators have a trait X_p that can be interpreted as a measure of the effort expended in catching prey. What is the coevolutionarily stable level of prey defense and predator effort, and what are the abundances of predator and prey at the coevolutionary equilibrium?

As before, the fitness functions in the prey and predators, respectively, are

$$W_v = 1 + r - (r/K)V - aP$$
$$W_p = 1 - \mu + abV$$

(29)

Now, as a simple example, assume that r is a decreasing function of X_v in order to indicate that there is a cost to prey defense. The benefit of this defense can be represented by supposing that a is also a decreasing function of X_v. Meanwhile, we can assume that a is an increasing function of X_p, the predator catching effect, and that the death rate μ is also an increasing function of X_p (because of the "cost" of X_p). Hence, a possible model for predator–prey coevolution is provided by the functions

$$W_v = 1 + r(X_v) - r(X_v)V/K - a(X_v,X_p)P$$

(30)

$$W_p = 1 - \mu(X_p) + ba(X_v,X_p)V$$

(31)

To analyze this model, we first differentiate W_v with respect to X_v and W_p with respect to X_p and obtain

$$\left(1 - \frac{V}{K}\right) \frac{dr(X_v)}{dX_v} - P \frac{\partial a(X_v,X_p)}{\partial X_v} = 0$$

(32)

$$-\frac{d\mu(X_p)}{dX_p} + bV \frac{\partial a(X_v,X_p)}{\partial X_p} = 0$$

(33)

Next, we set $W_v = 1$ and $W_p = 1$ and obtain

$$P(X_v,X_p) = \frac{r(X_v)}{a(X_v,X_p)} [1 - \frac{V(X_v,X_p)}{K}]$$

(34)

50

$$V(X_v,X_p) = \frac{\mu(X_p)}{ba(X_v,X_p)} \qquad (35)$$

Now, substituting for P and V into (32) and (33), we obtain a pair of equations for X_v and X_p themselves.

$$\frac{1}{r(X_v)} \frac{dr(X_v)}{dX_v} - \frac{1}{a(X_v,X_p)} \frac{\partial a(X_v,X_p)}{\partial X_v} = 0 \qquad (36)$$

$$\frac{-1}{\mu(X_p)} \frac{d\mu(X_p)}{dX_p} + \frac{1}{a(X_v,X_p)} \frac{\partial a(X_v,X_p)}{\partial X_p} = 0 \qquad (37)$$

The solution to these equations will give a level of prey defense X_v and predator effect X_p that are coevolutionarily stable, provided the solution represents a maximum, and not a minimum, for each fitness. The CSC abundances are then found from (34) and (35).

To use these equations to obtain a biological result, a submodel for the trade-offs must be introduced. If the predation rate can be written as a product of two factors,

$$a(X_v,X_p) = a_v(X_v)a_p(X_p) \qquad (38)$$

the equations become uncoupled, and can be written as

$$\frac{d \ln r(X_v)}{dX_v} - \frac{d \ln a_v(X_v)}{dX_v} = 0 \qquad (39)$$

$$-\frac{d \ln d(X_p)}{dX_p} + \frac{d \ln a_p(X_p)}{dX_p} = 0 \qquad (40)$$

Each of these can be solved graphically, and the conditions for an acceptable solution will depend on the details of the submodel that is chosen. A possible specific submodel for $a(X_v,X_p)$ is

$$a(X_v,X_p) = a_0(1 + \exp(-\alpha X_v))X_p^\beta \qquad (41)$$

where $X_v, X_p \geq 0$, and $\alpha, \beta > 0$. Also, we can take

$$r(X_v) = r_0 - \gamma X_v \qquad (42)$$

$$\mu(X_p) = \mu_0 + \delta X_p \qquad (43)$$

with $\gamma, \delta > 0$. The important point is that by using other choices of submodels, a model for predator–prey coevolution leading to a coevolutionarily stable community can be tailored to specific systems in which empirical studies are being done.

Two other topics in predator–prey coevolution concern the coevolution of the prey with one another caused by the presence of a common predation pressure. In Batesian mimicry, a palatable prey evolves to resemble an unpalatable species. Müllerian mimicry is the coevolution of unpalatable prey species to resemble one another. Theory for the evolution of mimicry has a venerable tradition tracing to Fisher (1930) and earlier (see Chapter 12 by Gilbert). Recent references include Nur (1970), Matessi and Cori (1972), Charlesworth and Charlesworth (1975), Huheey (1976), and Turner (1977, 1981). A related topic concerns coevolution of phenotypic differences among palatable prey, coevolution producing the result that experience by predators in catching one prey type does not enhance their ability to capture other prey species. This topic is known as the evolution of "aspect diversity" among prey. It is treated in Ricklefs and O'Rourke (1975), Endler (1978), and Levin and Segel (1982). The evolution of polymorphism in color within a species caused by predation from visual predators having a search image is known as "apostatic selection" (Cain and Sheppard, 1954; Clarke, 1969). This topic is also discussed in the papers concerned with aspect diversity cited earlier.

Coevolution between symbiotic partners

Symbiosis is a population interaction based on long-term physical contact between individuals of different species. Typically, an individual of one species permanently lives on, or in, a particular individual of another species. The relationship is usually asymmetric in the sense that a member of one species, usually the one that is physically bigger, can be identified as the "host" individual, and the other as the "guest." Traditionally, the symbiosis is considered to be parasitism if the guest exploits and harms the host, commensalism if the guest has little or no effect upon the host, and mutualism if both guest and host benefit from each other.

For examples of symbioses in marine environments see Vermeij (Chapter 14), Hedgpeth (1957), Wright (1973), Bloom (1975), Vance (1978), Losey (1978), and Osman and Haugsness (1981). Terrestrial examples of mutualisms are also accumulating: see Way (1963), Smith (1968), Springett (1968), Bentley (1977), Quinlan and Cherret (1978), Addicott (1979), Brown and Kodrick-Brown (1979), Janzen (Chapter 11), and Feinsinger (Chapter 13). In many of these examples, the mutualist modifies the interaction of the host with another species. The main questions are how mutualism can evolve, and whether a mutualistic association is ecologically stable.

For examples of parasitism, see Levin and Lenski (Chapter 5), Barrett (Chapter 7), Holmes (Chapter 8), and May and Anderson (Chapter 9). Also, Price (1980) has written a monograph on the population biology of parasites. General issues are whether evolution leads to a reduced virulence by the parasite toward its host and whether parasites maintain genetic polymorphism in the host population.

Because the individuals of different species are physically associated with one another for a long time, the survival of at least one member of the association depends upon the survival of the other member. Hence, the fitness of an individual depends on the *fitness* of its associate. This is a very important point. In the preceding examples, the fitness of an individual depended on the *number* of individuals of other species and not on the fitness itself of those individuals. But, with symbiosis, the fitness of an individual in a given species does depend on the fitness as well as upon the number of individuals of the other species.

Let us consider fitness functions for the parties in a symbiotic interaction (Roughgarden, 1975). Let upper case letters denote properties of the host, and lower case letters stand for the "guest." We suppose that an individual can potentially exist in either a solitary or associated state and that the survival and fertility is different for these states. To be in the associated state, a guest must find a host and the host must survive, otherwise the guest is returned to the solitary state. Similarly, to be in the associated state, a host must be found by a guest and the guest must continue to survive.

$$W_g = (pL_a)l_a m_a + (1 - pL_a)l_s m_s \qquad (44)$$

$$W_H = (Pl_a)L_a M_a + (1 - Pl_a)L_s M_s \qquad (45)$$

where p is the probability that a guest finds a host; L_a and l_a are the survivals of the host and guest, respectively, in an associated state; m_a and m_s are the guest fertilities in associated state and solitary state, respectively; P is the probability that a host is found by a guest; and M_a and M_s are the host fertilities while associated and solitary, respectively. The fertilities and survivals may be density dependent.

It seems natural to view symbiosis in terms of two issues simultaneously. First, to form an association, the guest must expend effort that decreases its fitness as a solitary individual in order to locate and then to enter the host. Also, the host may expend effort that decreases its fitness as a solitary individual either to repel the colonization attempts by potential guests if the guest is harmful to it or to solicit colonization if the guest is beneficial. Second, the guest must assess the degree to exploit the host. To lower the host's fitness sufficiently will cause the host's death and a forfeit of further benefit from the association, whereas not to exploit the host at all would lead to no benefit from the association to begin with. Similarly, the host must assess the extent to which it is advantageous to resist the pressure of a harmful guest once it has colonized, or to facilitate a beneficial guest.

Roughgarden (1975) explored these issues from the point of view of an evolving guest population, assuming the properties of the hosts could be taken as constant. Also, the formation of the association was viewed as occurring before the molding of the association and so these issues were treated separately, not simultaneously. The analysis was organized around the example of the damsel fish-sea anemone association originally studied in the Pacific by Verwey (1930). Roughgarden concluded that mutualism could only evolve in a restricted circumstance because of two considerations.

The potential host individuals must survive sufficiently well for a symbiotic association to form initially; otherwise, a potential guest might have a higher fitness by remaining unassociated with such a host. From this argument it follows that more species of obligate symbionts should have evolved to live on or in long-lived hosts than on or in short-lived hosts, everything else being equal.

The other consideration is that the extent to which a guest should forego exploitation of the host in order to increase the host's survival depends on how much that sacrifice yields in terms of an improvement to the host's survival. If the host already survives sufficiently well, or if the guest can have little impact on the host's survival, there may be little the guest can do to improve upon a host's survival. Thus, the evolution of mutualism by the guest toward the host requires that the host have an intermediate survivorship. It cannot be too low or otherwise the association will not form; it cannot be too high or otherwise there is no benefit to the guest of improving it still further.

If the guest evolves some moderation in its exploitation of the host, it might be called a "gentle parasite," provided the host nonetheless is harmed by the guest. For the guest to be a mutualist, it must actually increase the survivorship of the host above that where it is solitary. The guest is, of course, evolutionarily "unaware" of how well the host survives in the associated state as compared with the solitary state. The evolution of mutualism thus requires a lucky coincidence. The improvement to the host that the guest finds justified from its cost–benefit analysis must happen to be so great that the host finds itself actually aided by the guest. And once this should happen, the host should actively solicit the guest, and reciprocal mutualistic coadaptations can evolve. These considerations were applied to offer an explanation for variation among the relationship between several pairs of damsel fish and sea anemone species.

A problem with the Roughgarden (1975) analysis is that it is not a true coevolutionary analysis. It is a cost–benefit discussion from one party's point of view in a two-party interaction. A coevolutionary formulation would be as follows. Let x_c and x_m be indices of colonization

54

effort by the guests and of the degree of exploitation of the hosts by the guests. Similarly, let X_c and X_m be indices of the host's response to colonization by the guest and of the host's reaction to the presence of a guest that has colonized. Let g and H denote guest and host population sizes. Then the mathematical problem takes the form

$$\text{maximize } W_g\,(x_c, x_m, X_c, X_m, g, H\,) \tag{46}$$

with respect to both x_c and x_m, and set $W_g = 1$; and

$$\text{maximize } W_H\,(x_c, x_m, X_c, X_m, g, H) \tag{47}$$

with respect to both X_c and X_m, and set $W_H = 1$. This is a bivariate maximization in each species together with the requirement of ecological equilibrium. The coevolution of symbiotic populations in this way is unexplored, although the problem would be theoretically tractable if useful submodels can be found for the relationship between the effort indices and the parameters of the fitness function.

Up until now we have only considered individual selection. When the guest is physically much smaller than the host, as is almost a rule with parasites, then the host may contain a group of guests within it. In this context, group selection may become important as well.

Group selection is the genetic evolution of a population (such as a group of parasites within an individual host) through the differential extinction rates and through the differential production of groups that vary in their genetic composition. There are now two well-developed theoretical pictures of how group selection operates. In the first, the whole population is considered to be divided into many small local populations. These go extinct with a probability that depends on the composition of the local population. When a group goes extinct, its place is colonized by a fair sample of some neighboring local population. Also, there is a low level of interchange among all the local populations. Eshel (1972) has shown that it is theoretically feasible for group selection to cause the evolution of a gene whose carriers promote group survival even if such a gene is opposed by individual selection within each of the groups (see Roughgarden, 1979, pp. 283–292, for numerical examples of Eshel's model). Of course, if the gene both promotes group survival and also is favored by individual selection within the groups, evolution will occur even faster than with individual selection alone.

In the second picture of how group selection can work, the population is considered to be divided into small groups during most of the life cycle. All the groups emerge at some point for mating and subsequent redistribution into groups. The groups contribute differentially to the total population, depending on their composition. This approach has been pioneered by D. S. Wilson (1975b, 1977, 1979) who has shown that a gene that confers high group productivity can evolve even if it is neutral to, or opposed by, individual selection within the groups. Wilson (1980, Ch. 2), offers a summary of this work.

The main application of group selection ideas to symbiosis is on the problem of the evolution of reduced virulence by parasites toward their hosts. Day (1974) suggests there is a general evolutionary tendency for viral or bacterial disease to become less virulent, as seems to have happened with the myxoma virus that was introduced to wild rabbits in Australia (Fenner, 1971). The evolution of reduced virulence by the virus may have been accomplished in part through group selection according to the model by D. S. Wilson.

Conditions for the evolution of reduced or increased virulence by a parasite toward its host are derived by Anderson and May (1978). They begin with equations for the population dynamics of the host population, assuming that hosts exist in one of three states: susceptible, infected, and immune. They exhibit conditions whereby genes in the parasites that lower their virulence increase when rare and a comparable condition whereby parasites evolve higher virulence. The treatment is not coevolutionary; the hosts are assumed to be evolutionarily static during the time that the parasites evolve. Also, Levin and Pimentel (1981) have studied the evolution of virulence and focused on the connection between virulence and transmissibility. These studies are reviewed in more detail in Chapter 9 by May and Anderson.

There has also been an explicitly genetic approach to host–parasite coevolution using what are called gene-for-gene models. Flor (1955) found in flax (*Linum usitatissimum*) that there were dominant genes conferring resistance to particular genotypes of the rust (*Melampsora lini*). There is a one-to-one matching whereby a particular allele in flax confers resistance to a corresponding allele in rust. Person (1966) and Day (1974) report that such gene-for-gene systems are widespread among other plant species and their parasites (see also Chapter 7 by Barrett). Models for coevolution in gene-for-gene host–parasite systems have been treated by Mode (1958), Jayakar (1970), Yu (1972), Gillespie (1975), Clarke (1976), and Lewis (1981a). Most of these studies focus on the potential role of parasites in maintaining genetic polymorphism in the host, rather than on the effect of genetic changes on the ecological relationships of the host–parasite system.

Our final theoretical issue concerned with symbiosis is whether or not mutualism should be rare in nature because it is a destabilizing interaction within the community (May, 1973). The mathematical basis for the claim traces to a model of mutualism that is generated by reversing the signs of the competition coefficients in the Lotka-

Volterra competition equations. The fitness functions for this model are

$$W_1 = 1 + r_1 - (r_1/K_1)N_1 + \alpha_{12}(r_1/K_1)N_2$$ (48)
$$W_2 = 1 + r_2 - (r_2/K_2)N_2 + \alpha_{21}(r_2/K_2)N_1$$

The population dynamics are then given by

$$\Delta N_i = (W_i - 1)N_i, \qquad i = 1, 2$$ (49)

If the positive density dependence between the species exceeds the negative density dependence within the species, then coexistence may be impossible. In this case, both populations mutually reinforce one another and, mathematically, tend to infinite abundance in finite time. This issue may be artificial because the fitness function (48) seems to be a poor model for mutualism, because it does not take account of the physical closeness of the mutualists with one another (it is only density-dependent). Hence, it is impossible for mutualism to evolve with these fitness functions under ordinary natural selection. In contrast, natural selection can, under some condition, cause the evolution of mutualism with the fitness functions given by (44) and (45). The population dynamics of the mutualists based on (44) and (45), assuming density dependence in the survivorships and/or fertilities for both associated and solitary states, lead to abundances that are mathematically bounded and that pose no theoretical problem. Several recent papers investigating the stability of a mutualistic interaction include Vandermeer and Boucher (1978), Goh (1979), Travis and Post (1979), Hallam (1980), Heithaus et al. (1980), and Addicott (1981). There are also papers showing that species can have a positive effect on one another through indirect routes (Lawlor, 1979; Vandermeer, 1980).

PART 4.
COEVOLUTION AND COMMUNITY STRUCTURE

There are now some preliminary explorations of how coevolution may influence several interacting species, and possibly even an entire community. The main questions that have been examined are these: How does evolution in one species affect the abundance of that species and other species in the community? Are species groups themselves units of selection? What is the effect of coevolution on the stability of the whole community? Where and how fast does community coevolution take place? And what would be signs that a community was indeed coevolved?

57

Density-dependent selection acting on a solitary population leads (locally) to the highest equilibrium population size. This result means that for a solitary population the consequence of the spread of traits that confer a higher fitness is that the abundance of the population increases.

This result is not generally true in coevolution, although it is in special cases. When an evolving population interacts with other species, its abundance may or may not increase, depending on the nature and strength of all the ecological interactions of the population with other populations. For example, equations (25)–(27) in the predator–prey model show that evolution in the prey of traits conferring a higher carrying capacity K do not affect the equilibrium prey abundance. This result is a special case that has been treated more generally.

Consider first the situation where there is a trait X_s in each species that affects any, or all, of the parameters for the population dynamics of that species except the interaction coefficients with other species. For example, if species s is a prey population, the X_s might affect r_s and K_s but not the predation coefficient a.

$$W_s = W_s(X_s, N_1, \ldots, N_S), \quad s = 1, \ldots, S \tag{50}$$

At ecological equilibrium, we can implicitly define each equilibrium population size as a function of the value of the trait in a given species. For example, take Species 1: by differentiating $W_s = 1$, for all s, implicitly with respect to X_1, we have

$$\frac{\partial W_1}{\partial X_1} + \frac{\partial W_1}{\partial N_1}\frac{\partial N_1}{\partial X_1} + \cdots + \frac{\partial W_1}{\partial N_S}\frac{\partial N_S}{\partial X_1} = 0$$

$$0 \quad + \frac{\partial W_2}{\partial N_1}\frac{\partial N_1}{\partial X_1} + \cdots + \frac{\partial W_2}{\partial N_S}\frac{\partial N_S}{\partial X_1} = 0 \tag{51}$$

$$\vdots$$

$$0 \quad + \frac{\partial W_S}{\partial N_1}\frac{\partial N_1}{\partial X_1} + \cdots + \frac{\partial W_S}{\partial N_S}\frac{\partial N_S}{\partial X_1} = 0$$

We can then solve for $\partial N_1/\partial X_1$ and all the other $\partial N_s/\partial X_1$ using Cramer's rule. But, by inspection, we can see that if X_1 maximizes the fitness in Species 1, then $\partial W_1/\partial X_1$ is 0 and therefore all the $\partial N_i/\partial X_1$ are 0 also. Hence, the value of the trait in Species 1 that maximizes fitness in Species 1 produces either a local maximum or minimum to the abundance of every species, including Species 1 itself. Now every equation in (51) can be differentiated again to find the second derivatives. The sign of the second derivatives will tell us if the critical points are maxima or minima. In this way we can establish that Species 1 can maximize or actually *minimize* its abundance as a result of the spread

58

of traits that lead to fitness maximization within Species 1. Whether the spread of fitness-increasing traits in Species 1 increases or decreases the abundance of Species 1, depends on the entries in the matrix of derivatives $\partial W_s/\partial N_i$ in the equation above.

There is a condition based on the inverse of the matrix, $\partial W_s/\partial N_i$, which completely determines whether the evolution of fitness-increasing traits within a species maximizes or minimizes the abundance of that species (Levins, 1975; Roughgarden, 1977). This condition can be interpreted in terms of the concept of a "keystone" species.

A keystone species is a species whose removal leads to a still further loss of species from the community. For example, Paine (1966) found that in the absence of starfish, mussels overgrow and eventually exclude barnacles within a local study site. But starfish prefer to feed on mussels, so that in the presence of starfish, barnacles and mussels coexist. The starfish is a keystone species because its removal leaves a community that loses still more species.

There are two distinct methods whereby a species can become a keystone. In the first, the species is necessary for the existence of a positive equilibrium. Paine's example is of this type. Also, there are keystone competitors in this sense (see Roughgarden, 1979, pp. 546–548). What these examples have in common is that the resource arrangement without the keystone does not allow coexistence, but the dynamics are stabilizing in character. I term this type of keystone species a *positivity-enhancing keystone*. In the second method, the species is necessary to stabilize the system. Without the species, the remaining community may have a resource arrangement consistent with coexistence, but the nature and strength of the interactions do not combine to restore perturbations from this equilibrium. I term this type of keystone species a *stability-enhancing keystone*.

A mathematical example of a stability-enhancing keystone is provided by a model of the second species to follow a pioneer species in primary succession. Suppose the pioneer species, by affecting the substrate, makes the habitat better for itself and for the next arrival. If the first species has positive density dependence (it is autocatalytic), then it is unstable by itself. Yet it may stably coexist with a second species if that species brings sufficient negative density dependence to the system. The second species is thus a keystone species, but it is so by virtue of bringing stability to the dynamics. It is easy to construct more complicated examples using more species.

Now the surprising result is this: *if the evolution of fitness-increasing traits minimizes the abundance of a species, then that species is a stability-enhancing keystone.* As adaptive traits accumulate, the popu-

59

lation size of such a species continually declines, until extinction results. A stability-enhancing keystone is, in effect, a time bomb within a community. When it does go extinct, as a result of coevolution, then the remaining community will lose still more species. To the extent that coevolution structures communities, we should expect to see only positively-enhancing keystone species in nature. For a numerical example of a stability-enhancing keystone species evolving a minimum population size, see the pioneer–competitor example in Roughgarden (1979, pp. 470–471).

The preceding discussion concerns the evolution of traits that do not immediately affect the population interactions themselves. To consider traits that do mold the interaction coefficients, the fitness functions become functions of all the traits and all the population sizes,

$$W_s = W_s(X_1, \ldots, X_S, N_1, \ldots, N_S), \qquad s = 1, \ldots, S \tag{52}$$

If we proceed as before, we immediately find that, in general, $\partial N_i / \partial X_j$ is not 0 in the coevolutionarily stable community. With further analysis one can calculate the sign of $\partial N_i / \partial X_j$, which tells us whether the CSC value of X_j is below or above that which would maximize N_i. Thus, from the standpoint of population abundance, the coevolution of the interaction coefficients always leads to an adaptational mismatch in the sense that the values of the traits that produce a coevolutionarily stable community are virtually never those values that would lead to the highest abundance for any species in the community.

Groups of interacting species as units of selection

The preceding treatment pursues the idea that coevolution is caused by natural selection at the individual level in each of the interacting populations. But if there is a local grouping of members of different species with one another, the structure of the grouping may, in principle, influence the coevolution in ways that parallel the workings of group selection within a single species. D. S. Wilson (1980) has advanced this idea and has begun to develop it theoretically.

A good starting point is the "case of the nasty competitor" introduced by Roughgarden (1976). The trait of pure interspecific nastiness is where a member of one species interferes with a member of another species without directly benefiting from the act. Suppose the trait is in Species 1 and affects Species 2. Using the terminology of the competition equations, the trait X_1 influences $\alpha_{2,1}$ but not $\alpha_{1,2}$. But the fitness of a member of Species 1 depends only on $\alpha_{1,2}$ and not at all on $\alpha_{2,1}$. Hence, the trait is selectively neutral. Yet, if it did happen to spread, then the consequence would be to increase N_1 and to lower N_2. If the members of Species 1 occur in small local groups with members of Species 2, subsequently reassemble into a well-mixed, single-species population, and then redistribute at random into small local populations where both species occur, group selection within Species 1 may cause the evolution of pure interspecific nastiness. The reason is that groups where pure interspecific nastiness is in high frequency contribute a higher production to the population at the mixing phase than groups where interspecific nastiness is rare. Wilson (1980) also focuses on mutualism as possibly evolving in this way.

These ideas are very preliminary. What happens if both species evolve through group selection? That is, we need a true coevolutionary theory of group selection and not an analysis from the point of view of only one party. Do the group structures for both species need to coincide, or can they overlay one another in an uncorrelated mosaic? And so forth. The idea of coevolutionary group selection is promising and needs much more theoretical exploration.

Coevolution and community stability

May (1973) championed the view that complex communities were inherently more unstable than simple ones, a view that was counter to the conventional wisdom at that time. Simple agricultural systems were known often to have large fluctuations in abundance through time and were felt to be more vulnerable to perturbation, especially from pests, than complex natural communities. It seemed that a complex natural community would have checks and balances among the populations that would stabilize it to perturbation. May showed, using models, that it is very touchy to adjust the interactions among the members of a complex community so that the combined effects of the interactions are stabilizing; more often than not, the combined effects of the species interactions are actually destabilizing.

The question then becomes whether the overall consequence of coevolution is to produce communities whose species have just the right characteristics necessary to obtain a complex stable community. Based on the coevolutionary theory so far developed, the answer comes in two parts.

Coevolution itself often, perhaps even typically, destabilizes a community that was stable to begin with. Coevolution frequently leads to species extinction and shapes population interactions in ways that are less stable than the original condition. We have seen three examples where coevolution can lead to extinction. They are (1) in predator–prey coevolution, evolutionary evasion by the prey (i.e., evolution of the predation coefficient a to a value low enough that the condition for predator existence, $abK > \mu$ is violated); (2) in extinction of a com-

61

peting species during the coevolution of two competitors with asymmetrical competition; and (3) in extinction of stability-enhancing keystone species. Another example of coevolution destabilizing a population interaction is the evolution of a higher K by the prey in a predator–prey situation.

Nonetheless, because a frequent outcome of coevolution is the loss of species, the resulting community tends to be more stable than the original community simply because it is less diverse. Furthermore, one may conjecture that the remaining species in the less diverse community have weaker interactions with one another than an arbitrary community of equal diversity. Coevolution may, so to speak, prune the strongly interacting species from the original community, leaving a community that is both less diverse and has weak interactions among most remaining members. At this time, the idea that coevolution prunes the strongly interacting species from a community is purely conjecture; what is clear is that, in coevolutionary models, the extinction of species is a frequent outcome and that the resulting less diverse community will, for at least this reason alone, generally be more stable than the original noncoevolved community. Thus, it is not by "molding" the interactions, but by causing extinctions, that coevolution may ultimately generate stable communities.

The generalized taxon cycle

E. O. Wilson (1961) advanced the idea that there is a cyclic process in the ant fauna of South Pacific islands such that extinctions occur within an island fauna followed by invasions. The sequence of extinction and invasion leads to a turnover of species identities while preserving the total species number in the fauna at a roughly constant value. It is evident in Wilson's discussion that the extinction results from an interaction between existing members of the ant fauna. According to Wilson, the extinction is not the immediate population-dynamic consequence of the successful invasion of a new form into the fauna; rather, the invasion is made possible by the extinction, together with the withdrawal of resident species from marginal habitat as they shift toward use of preferred forest habitat.

Ricklefs and Cox (1972) have suggested that a type of taxon cycle occurs among the birds of the West Indies. The overall picture is complicated because some bird groups may participate in a cycle and others may not. Pregill and Olson (1981) have expressed doubts about the reality of the taxon cycle in West Indian birds and suggest that long-term changes in the environment are the primary cause of extinctions.

The *Anolis* lizard population on the small islands of the eastern

Caribbean also appear to undergo a cyclic process, as discussed in more detail in Chapter 17 (see also Roughgarden et al., 1982).

The theory of coevolution predicts that there are many situations in which extinction is the consequence of the coevolutionary process, as mentioned previously. If extinctions are followed by invasions, then cycles consisting of alternations between extinction during coevolution followed by invasion might be quite common. I term such a cycle a "generalized taxon cycle."

The idea that frequent extinctions result from coevolution is not inconsistent with a major role being played by climate changes during geologic time as discussed by Pregill and Olson (1981). Climatic changes may effect an extinction that has been "set up" by biotic interactions, including coevolution. There can be a "biological targeting" of the effects of environmental change. For example, in the taxon cycle that seems to be occurring on St. Maarten in the eastern Caribbean, the smaller species *Anolis wattsi* is expected to become extinct in the future. The coup de grace will probably be a 10- to 20-year run of comparatively dry rainy seasons during which the central hills where *A. wattsi* lives will approach the xeric state of the sea-level habitat in which *A. wattsi* is excluded by its competitor, *A. gingivinus*. Thus, the connection between environmental change during geologic time and the extinction of species in the fossil record may be provided by an understanding of community processes, including the role of coevolution in promoting species extinction.

Possible limitation of effects of coevolution to systems of low diversity

The coevolutionary theory of two competitors has been extended, numerically, to more than two competitors (Rummel and Roughgarden, in preparation). From this work it is clear that the time needed for a set of competitors to reach coevolutionary equilibrium increases very rapidly with the number of coevolving species.

The reason that coevolution works slowly in high diversity competition systems is twofold. First, any shift in niche position that decreases competition with one species increases the competition with another species, thereby reducing the net direct selective advantage to such a shift. Second, species influence others through long and indirect pathways, pathways that are longer and whose effects are weaker in high-diversity systems. Weak direct and indirect selection pressures require a long time to produce results.

The relationship between species diversity and the time required for a coevolutionary equilibrium to be attained has not been investigated in other model systems. Nonetheless, it seems reasonable to conjecture that high-diversity communities will generally equilibrate more slowly than comparable low-diversity systems, because the reasons that underlie this relationship in competition communities are not special to the competitive interaction.

Similarly, Case (1982) has pointed out using a two-species coevolutionary competition model that if extinction does not occur the coevolved niche separations are large. They are so large that invasion by other species can readily occur. He suggests that coevolved systems will only be found in locations that are rather inaccessible to invasion, such as distant oceanic islands.

Thus, low diversity and remoteness promote the likelihood that coevolution has been an important contributing cause to the formation of community structure. The possible importance of coevolution in diverse, but very permanent, communities is presently unexplored. However, diverse communities may contain relatively isolated subsystems of low diversity within which coevolution may occur very quickly.

PART 5. CONCLUSION

This chapter testifies to the existence of the beginnings of a comprehensive theory for coevolution among interacting populations. As I see it, the main goal for future work in coevolutionary theory is to develop and to test models that are tailor-made for the particular coevolved systems in which empirical studies can be feasibly done. There is a glaring shortage, even absence, of good models for many kinds of coevolutionary situations, including the plant–herbivore and plant–pollinator interactions. The predator–prey models seem artificial to me and should be replaced with more system-specific models, like those for arthropod predation developed by Hassell (1978). The models based in epidemiology used by Gillespie (1975), May and Anderson (1978), and Levin and Pimentel (1981) also seem well posed. Models for the coevolution of competitors are appropriate for some systems but have unknown generality. We also need models that are natural for the population structure of space-limited marine populations having sessile adults and pelagic larvae. With such tailor-made models, we can increasingly develop coevolutionary predictions that are concrete enough to be rigorously tested under field conditions.

CHAPTER 4

PHYLOGENETIC ASPECTS
OF COEVOLUTION

Charles Mitter and Daniel R. Brooks

INTRODUCTION

Phylogenetics is the discipline that attempts to reconstruct the genealogical relationships among taxa and the sequence of origin of their distinguishing features. In this chapter we will address the way in which phylogenetic methods might be used to study the historical origin of contemporary species interactions and the characteristics that govern them. We will treat two general questions. The first is that of when and how any given association became established. To what extent, we may ask, does a set of currently interacting species represent the descendants of similarly associated ancestors? The second question is that of the consequences of "association by descent": To the extent that interacting lineages have been associated through time, how and to what degree have they influenced each other's evolution? This is the question of "coevolution" as defined in the introduction to this volume.

The first of these issues, that of where associations come from, has received the greater share of attention from systematists. It is the one for which more phylogenetic evidence exists, and it will receive the more detailed treatment here. As relatively sophisticated phylogenetic analyses become more widespread, however, systematics should make an increasing contribution to the study of coevolution.

THE ORIGIN OF ASSOCIATIONS: "ASSOCIATION
BY DESCENT" VERSUS "COLONIZATION"

The notion of association by descent is an old one. A number of early workers in parasitology remarked upon apparent correspondences between the relationships of parasites and their hosts. Kellogg (1896) suggested that avian biting lice gave evidence of relationships among their hosts:

The occurrence of a parasitic species common to European and American birds, which is not an infrequent matter, must have another explanation than any yet suggested. This explanation, I believe is, for many of the instances, that the parasitic species has persisted unchanged from the common ancestor of the two or more now distinct but closely-allied bird species. [p. 51]

Observations of this kind, many of which are summarized by Metcalf (1929), were frequent enough that Eichler (1948) made them the subject of "Fahrenholz's Rule": "the natural classification of some groups of parasites corresponds with that of their hosts." Eichler accounted for this generalization by the theory that "the ancestors of extant parasites must have been parasites of the ancestors of extant hosts, so that the evolution of hosts and parasites has been in correspondence."

How might this assertion be tested by phylogenetic study? We will treat the problem as one of explaining the origin of new host associations during parasite phylogenesis. (It is also possible to approach the subject by tracing the acquisition of parasites during host phylogeny; see Brooks, 1981; Brooks and Mitter, 1983). Accounting for the origin of organisms' features is central to phylogenetics, and it is thus appropriate to analyze ecological associations by logic analogous to that used in phylogenetic reconstruction of character evolution. Before we do this, however, we will review briefly some salient features of phylogenetic methodology.

A précis of phylogenetics

Evolutionary relationships among noninterbreeding entities (e.g., species) can be represented as a phylogenetic tree or cladogram (see, for example, Figure 1). The terminals at the ends of branches (e.g., the *Enterobius* species in the figure) represent taxa on which character data have been gathered. The internal nodes or branch points may be interpreted as most recent common ancestors of all forms lying on paths connecting to and "above" (i.e., closer to a terminal than) them. The line segments thus represent the succession of ancestral and descendant forms within single lineages. Any group of taxa composed of all the descendants of a given ancestor, and only those descendants, is *monophyletic* in the sense of Hennig (1966); the members of a monophyletic group thus share a common ancestor not possessed by any taxon outside that group. For convenience we will often refer to a group by equating it with either its most recent common ancestor or the lineage giving rise to it. Thus, in Figure 1, we may denote the

Enterobius species   Primate hosts

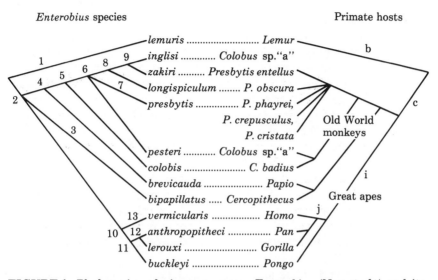

FIGURE 1. Phylogenies of pinworm genus *Enterobius* (Nematoda) and its primate hosts according to Brooks and Glen (1982). Primate phylogeny follows Schwartz et al. (1978), cited in Wiley (1981).

group composed of *Enterobius anthropopitheci*, *E. lerouxi*, and *E. buckleyi* as group 11, or that containing just *Homo*, *Gorilla*, and *Pan* as group j.

A monophyletic group comes by definition into existence when the ancestor of the larger group to which it belongs undergoes speciation. Any two monophyletic groups that together make up a larger one are called sister groups and are, by definition, of equal age. In Figure 1, groups 1 and 2 are sister groups, and both are older than any of the groups included within group 2.

A widely adopted approach to the construction of phylogenetic trees follows the "phylogenetic systematics" of Hennig (1966). We will presuppose a set of homologous characters, general descriptors (e.g., number of cervical vertebrae), whose character states or particular conditions (e.g., seven, the number of cervical vertebrae found in mammals) have been scored for each taxon under study. Under Hennigian analysis, one seeks the most parsimonious genealogical "explanation" for the distribution of character states over taxa, that is, the phylogenetic tree requiring postulation of the fewest character state changes. The smallest conceivable number of changes for a specified character is clearly just one less than the number of states, but this minimum will only be realized if all occurrences of each state can be accounted for by a single origin. If a particular state (e.g., the possession of seven cervical vertebrae) is found in all and only the members of some putative monophyletic group (e.g., the animals currently classified as "mammals"), we need invoke just a single origin of that state, in an ancestor unique to that group. In contrast, if a state (e.g., homeothermy) occurs only in widely separated parts of a genealogy (e.g., in birds and mammals, among the amniotes), we may be required to postulate more than one origin for it; such states are said to show low *consistency* (Kluge and Farris, 1969) on the specified phylogeny. Several proposed methods assign ancestral states to each monophyletic group so as to minimize the total number of character changes. Fitch's (1971) method assumes that any state may be transformed directly into any other state. Farris' (1970) method uses a predefined "transformation series," that is, a rule that postulates an evolutionary ordering of the character states. Thus, the transformation series a↔b↔c stipulates that the character cannot be transformed directly from state a to state c, without passing through b. Both of these methods produce an estimated historical ordering of character states during evolution (Mickevich, 1978, 1981).

Mickevich (1981) has argued that a cladogram corroborates a particular transformation series if states adjacent in that series occur in taxa that are adjacent on the cladogram. (The best-fitting transformation series under this "nearest-neighbor" criterion also gives maximal consistency under the Farris procedure.) Theories about the nature of character transformation can thus be evaluated by the fit to cladograms of the transformation series deduced from them (see detailed treatment in Mickevich, 1981).

The parsimony criterion by itself does not place restrictions on which state is most ancestral (see Farris, 1970, 1979), and thus additional criteria (e.g., concordance with stratigraphic position) must be invoked to fix the "roots" of genealogies and character histories once their branching forms have been established (see, for example, Lundberg, 1972). Finally, we should point out that some workers have rejected the application of philosophical parsimony to phylogenetic inference and proposed other approaches (see, for example, references in Mitter, 1980).

Association by descent

The preceding discussion implies that a theory about the evolution of parasites' host associations must make a prediction about the order of origin of particular habits. It is appropriate to point out the close formal resemblance between the issue of "association by descent" versus "colonization" in historical ecology, and that of "vicariance" versus "dispersal" explanations for disjunct distributions in biogeography.

In biogeography, the geographical "disjunction" between related forms may have arisen simultaneously with (and as a result of) the separation between the areas they occupy, or later, as a result of colonization. Similarly, related parasites may occupy different hosts because they speciated in concert with the hosts or because the parasite of one became transferred to ("colonized") the other. Our discussion will therefore draw heavily on the recent literature of phylogenetic biogeography, particularly Mickevich (1981; also Nelson and Platnick, 1981). However, although there is agreement on broad principles, there is as yet no consensus on a general method for specifying the systematic consequences of "disjunction" theories.

Under the hypothesis of association by descent, origin of new host associations of parasites occurs solely through divergence of ancestral hosts into daughter species. The ancestral host of any monophyletic group of parasites must therefore have been the host ancestor giving rise to all hosts in which those parasites occur. In that case, the transformation series for the host association "character" of the parasites should look like the host phylogeny, with the addition of a state corresponding to the ancestor of each host group (see Nelson, 1974). For example, in Figure 6A, the transformation series $(1 \to 2 \to 3)$ expresses the history of host association in the lineage leading to the parasite of *Caiman* (Cai), state (1) being the common ancestor [Croc, All, Cai] of the three crocodilian genera, state (2) the common ancestor [All, Cai] of *Alligator* and *Caiman*, and state (3) *Caiman* [Cai]. Thus, the minimum number of host association changes that must have occurred is one less than the number of states. However, when ancestral host-association states are assigned to the nodes (common ancestors) in a parasite cladogram under the hypothesis of mutual descent of hosts and parasites, the number of host association changes that must be postulated will often be greater than the minimum that would be required if the host and parasite cladograms had been perfectly congruent. The ratio of the latter quantity to the former, called the consistency index (Kluge and Farris, 1969), will be higher the greater the correspondence between the ordering of hosts on the parasite phylogeny and their relative age according to the host phylogeny.

We will illustrate the approach just outlined using the genus *Enterobius*, a set of primate-infesting pinworms (Nematoda). Figure 1 depicts a phylogeny for these parasites, based on analysis of their morphology (Brooks and Glen, 1982), along with a widely accepted view of the genealogy of their hosts. Let us examine the distribution of several hosts over the parasite phylogeny for compatibility with an "association by descent" explanation. From the host cladogram we would predict, for example, that any event isolating a pinworm lineage on *Lemur* should occur before disjunctions separating parasites on individual genera of higher primates, because the latter did not exist until after the origin of *Lemur*. This expectation is in accord with the occurrence of the oldest *Enterobius* species in *Lemur*.

Within group 10, association by descent predicts that the lineage isolated on *Pongo* should be older than the one on *Homo* and that isolation on the latter should precede restriction to either *Gorilla* or *Pan*. To reconcile this expectation with the observed "reversal" of position between *Pongo* and *Homo*, we must postulate some history like that in Figure 2, which represents inferred ancestral hosts for each parasite group as the collection of all that host's descendants (Nelson, 1974). The independent derivation of parasite groups 12 and 13 from forms living in the same ancestral host could be explained by postulating, for example, that an ancestor of group 10 underwent a divergence event not affecting host lineage i, producing two sister parasites (10′ and 10″ in Figure 2) infesting i. Both diverged later along with group i, but representatives of group 10′ are missing from *Pongo* (as well as from *Pan* and *Gorilla*) due to extinction, failure to establish, inadequate sampling, etc., and group 10″ is absent for similar reasons from *Homo* (see discussion in Nelson and Platnick, 1981). Thus, if the phylogenies of the parasites and their hosts had been perfectly congruent, there would have been six changes in host association of the parasites, corresponding to the six line segments (internodes) in the phylogeny of the four hominoid genera from their common ancestor (Figure 1). Under the assumption of mutual descent, the *actual* phylogenies of the hosts and their parasites require seven changes in host association, described in Figure 2 by the solid internodes emerging from the two nodes marked [G, Pa, H, Po]. Thus, the consistency index is 6/7.

Although specific points of conflict can be found in the *Enterobius* example, a high level of overall agreement with the predictions under association by descent seems evident. It is important, however, to ask whether these observations might be accounted for equally well by a colonization model under which the host group diversifies completely before its members are sequentially invaded, with subsequent divergence, by the parasites. How might we choose among alternative possible histories for such colonization? In this case the states of the "host association" character are presumed to give rise directly to one another, so we could simply optimize this character according to the Fitch criterion. For group 10 of Figure 2 there are several equally good solutions, each requiring the minimum of three disjunction events. For example, the lineage ancestral to group 10 could have lived on *Pan* and then colonized in succession *Homo, Pongo,* and *Gorilla,* or it could

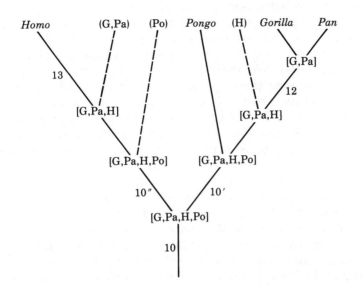

FIGURE 2. Hypothetical history for hominoid pinworms (group 10) of Figure 1, compatible with association by descent. Dashed lines represent lineages failing to establish in, or extinct from, the hosts in parentheses. Host ancestor inhabited by each parasite ancestor is represented by brackets containing extant hosts descended from that ancestral host. H, *Homo*; G, *Gorilla*; Pa, *Pan*; Po, *Pongo*.

equally well have started on *Pongo*, colonized *Homo*, and then given rise to a colonist of *Pan*, which subsequently colonized *Gorilla*.

Because some such colonization history can always be found that will maximize fit to the observations, how can we ever reject such an explanation? Association by descent accounts for disjunctions by a single principle, host phylogeny, whose consequences for the distribution of host associations can be specified and then tested. By contrast, a colonization hypothesis chosen simply to maximize fit to the parasite phylogeny is purely ad hoc. Colonization hypotheses are not inherently ad hoc, but to be maximally testable they too must make independent predictions about disjunction sequences, based on some property of the hosts such as geographical proximity or ecological similarity (see Mickevich, 1981, for extended discussion).

For most of the examples to be discussed, it is difficult at present to propose concrete alternatives to association by descent. The information required to construct such models, which would have to come

71

from studies of life histories and ecological interactions, is harder to obtain than the morphological data on which a phylogeny can be based. Thus, we will generally be limited here to evaluating association by descent hypotheses in isolation, comparing "expected" disjunction sequences (derived from host phylogenies) to observed ones ("phylogeny-like" sequences fitting the data best; see Figure 6; also Mickevich, 1981). We shall assess only qualitatively whether the agreement between these seems too close for coincidence; there has been no resolution of the issue of how (or even whether) sampling distributions for either transformation series or cladograms should be constructed, or what a reasonable null hypothesis in biogeography/coevolution might be (see, for example, Farris, 1981; Simberloff et al., 1981).

In fitting association by descent to our example so far, we have accounted for apparent deviations by assuming "extra" host divergences not supported by observation. It seems plausible that both mutual descent and colonization may have contributed to any observed set of associations, even when the overall pattern suggests strong agreement with one or the other mode of explanation. For example, it seems possible that the occurrence of *Enterobius inglisi* in a species of *Colobus* (Figure 1) might represent a host shift from *Presbytis*, because under mutual descent this association requires several "extra" disjunctions. There would be little point in abandoning a broadly corroborated mode of explanation just to achieve better fit in particular instances unless independent support could be found for the alternative theory. In zoos, however, pinworms are able to invade hosts not used in the wild (Brooks and Glen, 1982); the *Colobus* species is sympatric with *Presbytis*; and cooccurrence of two species (i.e., *inglisi* and *pesteri*) in the same host is otherwise unknown in *Enterobius*. Taken together these facts suggest that the recorded association of *E. inglisi* and *Colobus* is a recent one that may not even represent the typical host of this parasite. These partners are, however, descended from lineages that probably were associated for a significant fraction of the joint history of the groups to which they belong, i.e., up to the time of divergence between *Presbytis* and *Colobus*. Thus, mutual descent and colonization are not mutually exclusive; rather, we must inquire as to the relative contribution of each to the history of a parasite–host assemblage.

The example above raises other currently unresolved issues. First, how do we regard members of the host group that harbor no parasites? Thus, gibbons (*Hylobates* spp.) are generally regarded as the sister group of the rest of the great apes, but *Enterobius* has not been recorded from them. It can be argued that because the state "*Hylobates*" has not been observed for the pinworm host-association "character," a transformation theory for that character need no more concern itself

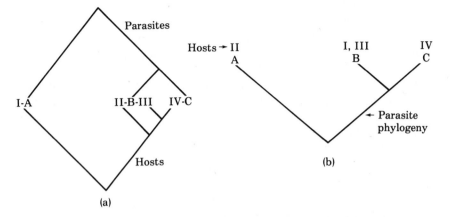

FIGURE 3. Hypothetical examples with multiple hosts for some parasites. (a) Pattern fully compatible with association by descent. Parasite ancestor (A,B) fails to differentiate when host ancestor (II, III, IV) splits to give II and (III, IV). Subsequent isolation splitting IV from III is accompanied by parasite speciation, isolating C on IV, B on II and III. (b) Same phylogenies as in a, but distribution of parasite over hosts not readily explained by continuous mutual descent.

with this than with any other nonobservation. Or it can be argued that the origination and subsequent divergence of a pinworm lineage restricted to the ancestor of great apes including *Hylobates* is required under association by descent and that any host transfer theory, say, that would place *Hylobates* at the end of instead of the middle of the transformation sequence, should gain credence as a result.

Second, the same parasite is frequently found in more than one host, as is true for *E. presbytis*. The general problem of how such observations are to be accommodated in disjunction theories is unsolved. Occupation of multiple phylogenetically adjacent hosts (not necessarily sister groups; see Hennig, 1966), resulting from host divergence without concomitant parasite speciation (see Figure 3a), seems fully compatible with mutual descent. Occupation of cladistically distant hosts by the same parasite, in contrast (Figure 3b), seems more difficult to explain by this process.

EVIDENCE REGARDING ASSOCIATION BY DESCENT

In this section we will review some of the systematic evidence that bears on the issue of how ecological associations typically originate. We will use a variety of examples to address two major questions: (1)

73

At what level (if any) in the genealogical hierarchies of associates is each of the two modes of origin most evident? (2) How do these patterns vary among different types of association?

For very few groups on which there is substantial ecological information have explicit phylogenetic trees been proposed, and for fewer cases still are there phylogenetic trees for both associated lineages. Moreover, it is frequently not possible to determine from the information supplied just how a published genealogy was derived. We have tried to choose cases that appear to have a reasonable grounding in morphological evidence, excluding, for example, those in which parasite phylogenies have been based partly on extrapolation from their hosts' relationships; in the examples below, we shall generally take the authors' taxonomic conclusions as given and shall merely examine their consequences for the questions raised above.

Animals living in, on, or with other animals

Some recent phylogenetic studies of animal host–parasite associations support the notion that these generally have long histories. In the *Enterobius* pinworms, as we have seen, there is good agreement at the highest taxonomic levels (i.e., involving the earliest divergences) with the expectation from mutual descent, but several points of departure within these larger groups. These include the "reversal" of position between parasites in *Homo* and *Pongo*, the possible "colonization" of *Colobus* sp. "a" by *E. inglisi*, and the placement of *E. brevicauda* with group 5, which is inconsistent with the requirement that any disjunction between *Cercopithecus* and *Papio* not precede the separation of these from the other old-world monkeys (Figure 1). But these are minor rearrangements, involving taxa relatively nearby on the tree; there is a strong overall association between the position of a host in the host phylogeny and its position in the disjunction rule that would fit best the parasite phylogeny. Given the number of taxa involved, the agreement seems too good to be accidental, and this case may be the best-documented instance of long-term mutual descent.

A phylogenetic analysis of digenean trematodes of the subfamily Acanthostominae (Brooks, 1980b) presents more complexity. A comparison with the vertebrate phylogeny shown in Figure 5 suggests that the Acanthostominae might have developed contemporaneously with their hosts, because the first split in both basal lineages (2 and 4) separates a teleost-inhabiting line from one in crocodilians. This hypothesis requires us to suppose, however, that the two trematode lineages have both diverged with lower vertebrates and yet left no descendants in any of the hosts that lie phylogenetically between crocodilians and perciform fishes (Figure 5). Moreover, the occurrence of group 12 in fishes would necessitate postulation of extra disjunctions to account for the restriction of earlier-diverging lineages 7 and 9

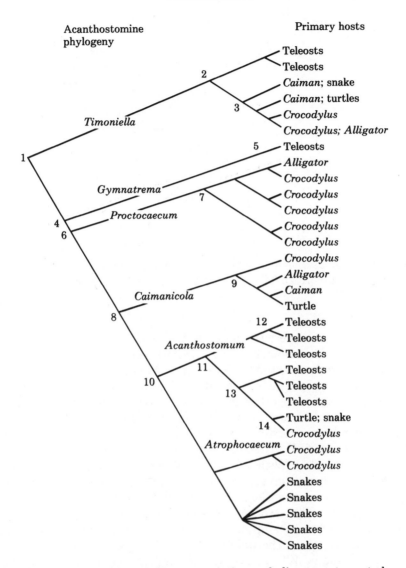

FIGURE 4. Phylogeny and host associations of digenean trematodes of the subfamily Acanthostominae (Cryptogonimidae), simplified from Brooks (1980b).

to crocodilians. Because all of the hosts of the acanthostomines are either fish or piscivores (Brooks, 1980b) or both, making parasite transfers among them plausible, it seems reasonable to suppose that their occupation at least of the distantly related major groups of hosts

75

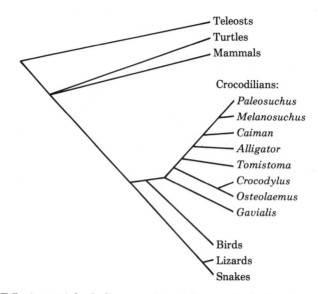

FIGURE 5. A partial phylogeny of vertebrates, including hosts of Acanthostominae (Figure 4), following Romer (1966), Sill (1967), and Neill (1971).

involved host transfer. If we divide the hosts into "teleosts" versus "reptiles" and assume that the nearest relatives of the acanthostomines, like the great majority of species in the family to which they belong, occur in fish, then the most parsimonious scheme of host transfer would show independent invasions of crocodilians by groups 3 and 6. Within group 11, we could either regard groups 12 and 15 as showing independent reversions to a fish-inhabiting habit or regard group 14 as an independent invasion of reptiles descended from a fish-parasitizing ancestor (11). With this part of ecological history provisionally established, we may investigate the possibility that *within* the major invasions, new associations have arisen predominantly by descent. Within group 3, the associations with turtles and snakes, like similar ones in *Caimanicola* and *Acanthostomum*, almost surely represent host transfers, given the wide phylogenetic separation of these hosts from crocodilians. What about the possibility that group 3 has otherwise undergone mutual descent with its predominant and presumably original host group, crocodilians? If we ignore the several host genera not harboring group 3, we can depict the disjunction sequence predicted from the generally accepted host phylogeny as in Figure 6A. Because host ancestors cannot reappear after splitting and because the oldest parasite species occurs in the youngest host, all three parasite ancestors must have lived in the oldest host ancestor and must have undergone subsequent mutual descent with the host group independently. Because a different, but unacceptable, host phylogeny (e.g., Figure 6B) would give much better fit, we may be justifiably skeptical about the notion of association by descent with crocodilians for group 3. By contrast, for group 6, which we shall interpret as representing a single invasion of crocodilians (with various derivatives colonizing other hosts), the best "association-by-descent" transformation series is the one corresponding to host relationships, requiring 11 disjunctions as opposed to 13 under the next-best possibility. Thus, although a much more formal and rigorous treatment is desirable, the data on acanthostomines would appear to rule out a history of continuous association in general with their hosts but offer limited support for mutual descent of particular subgroups of hosts and parasites.

Other cases involving internal parasites of vertebrates (see, for example, Brooks, 1977, 1978b, 1979a; Brooks et al., 1981) fall into the broad categories suggested by these examples. Long-term mutual descent seems to be a reality, as evidenced by groups like *Enterobius*, in which the order of "host disjunctions" corresponds well to host phylogeny. But there is strong evidence for host transfer in many cases, namely those, like some acanthostomines, requiring large numbers of ad hoc assertions under association by descent and in which ecological factors such as cooccurrence in the same habitat appear to be better explanations of the origin of host associations. Finally, there are some cases in which the data are compatible with association by descent, but to a degree ambiguous enough that more detailed analyses, with rigorous investigation of plausible alternatives, will be required before more definite statements can be made about them.

One approach to the resolution of "ambiguous" cases, urged by a number of earlier workers (e.g., Szidat, 1956) and elaborated by Brooks

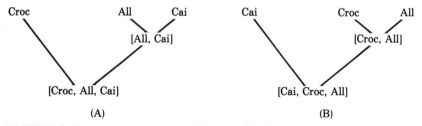

FIGURE 6. Two possible "mutual-descent-like" host disjunction rules for group 3 of Acanthostominae (Figure 4). (A) Cladogram that corresponds to actual relationships among hosts; (B) cladogram that does not, but fits better to parasite phylogeny. Croc, *Crocodylus*; All, *Alligator*; Cai, *Caiman*. For interpretation of brackets, see Figure 2.

(1981), is to broaden the inquiry to the entire complex of different parasite groups associated with the same hosts. The sum of these independent cases might suggest that the parasite fauna as a whole has developed along with the hosts and might suggest associated descent of individual groups that in themselves provide but weak evidence for it. Brooks (1981) concluded from such an analysis that the trematode fauna of crocodilians, including the groups discussed above, shows a general pattern of mutual descent.

We have concentrated on recent, explicitly phylogenetic studies because we consider these to provide the most definite evidence. However, many earlier workers, investigating a wide diversity of groups, have addressed these issues. Indeed, an elaborate series of "rules" for host–parasite evolution has developed. Some of the most important of these rules are as follows:

1. Fahrenholz's Rule (see Eichler, 1941a,b; Stammer, 1957; Cameron, 1964; Dogiel, 1964; Hennig, 1966; Ashlock, 1974): *Parasite phylogeny mirrors host phylogeny.*
2. Szidat's Rule (Szidat, 1956, 1960a,b): *The more primitive the host, the more primitive the parasites it harbors.*
3. Manter's Rules (see Manter, 1955, 1966; Inglis, 1971): (a) *Parasites evolve more slowly than their hosts.* (b) *The longer the association with a host group, the more pronounced the specificity exhibited by the parasite group.* (c) *If the same or two closely related species of host exhibit a disjunct distribution and possess similar faunas, the areas in which the hosts occur must have been contiguous at a past time.*
4. Eichler's Rule (see Eichler, 1941a,b, 1948; Inglis, 1971): *The more genera of parasites a host harbors, the larger* [i.e., the more speciose] *the systematic group to which the host belongs.*

Detailed reviews by Eichler (1948), Stammer (1957), and others drew the conclusion that conformity to Fahrenholz's Rule was widespread. However, most of the evidence cited for this generalization consisted only of the restriction of particular taxa of symbionts to particular taxa of hosts, without consideration of phylogenetic relationships within or among groups. To be sure, such correlations are a necessary consequence of mutual descent, and repeated failure to find them, especially when coupled with evidence for the importance of ecological factors in the evolution of host selection, would suggest a predominant role for host transfer. Thus, extensive studies on the gregarine sporozoans inhabiting various invertebrates (Stammer, 1957) showed that, whereas most of the species are restricted to more or less related host species, most gregarine genera are scattered across several host families, orders, or even classes. Supporting evidence for the

78

prevalence of host transfers in these symbionts is provided by the fact that encounters among different host species frequently result in "straying" of gregarines into the "wrong" hosts. A similar pattern obtains for many mites parasitic or commensal on insects, mammals, and birds (Stammer, 1957; but see Regenfuss, 1967). In the case of syringophilid feather mites, Kethley and Johnston (1975) propose that the variable most strongly predicting host associations among related mites is not bird taxonomy but quill size.

In contrast to the examples just cited, there are many instances in which pairwise associations of host and symbiont taxa are strongly evident. A widely cited case is that of the wood-digesting flagellates inhabiting the guts of lower termites, reviewed by Kirby (1937; see also Honigberg, 1970). Some groups of flagellates, for example, *Trichonympha*, are very widespread among the termites, with a distinct species group occurring even in the primitive wood roach *Cryptocercus*. Others, however, are restricted to particular termite groups. For example, the well-defined subfamily Pyrsonymphinae is found only in *Reticulotermes*. Termite flagellates are passed between host individuals by proctodeal feeding and seem incapable of forming resistant cysts. Coupled with the usual extreme social isolation of individual termite colonies, this makes the prospect of transfer among termite species in nature seem remote, although such transfers can be carried out artificially. Thus, many authors have concluded that present-day associations represent continuous mutual descent since the origin of blattoid insects, with subsequent loss of protozoa from higher roaches and termites.

Pairwise associations of particular host and symbiont groups by themselves cannot, however, be taken as strong evidence for association by descent, without more detailed examination of cladistic sequences either within or among those pairs (see Hennig, 1966). Such correlations might reflect only a tendency for host transfers to occur among near neighbors on the host phylogeny (Figure 7). Phylogenetic analyses of the many cases similar to the one just discussed (e.g., Rühm, 1956) would seem a fruitful area for future coevolutionary studies.

Szidat (1956) considered Szidat's rule a necessary long-term corollary to Fahrenholz's rule and described several cases of parallels between host and parasite groups in relative "primitiveness" (i.e., branching order as inferred from a presumed transformation series of one or more characters). For example, 11 subfamilies of paramphistonid trematodes, ordered according to the position of the testes, corresponded to the presumed sequence of origin of their fish, am-

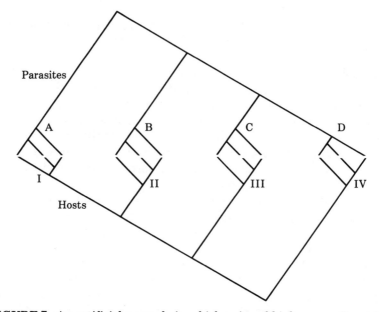

FIGURE 7. An artificial example in which pairs of higher parasite and host taxa are strongly associated. However, phylogenies among and within these taxa rule out association by descent.

phibian, sauropsid, and mammalian hosts (Szidat, 1939, cited in Eichler, 1948). A more recent example of such correspondence is that described by Radovsky (1967), involving 15 genera of macronysiid and laelapid mites and the bat families on which they live. It will be of great interest to see whether the conclusion of associated descent, which is strongly suggested by such cases, is supported by phylogenetic analysis. Some groups postulated in the literature to show mutual descent have nevertheless been described, somewhat paradoxically, as failing to obey Szidat's rule. For example, Kirby (1937) stated that although it seemed clear that many symbiotic flagellates had evolved together with their termite hosts, there was no overall tendency for the primitiveness of a flagellate fauna to reflect that of its host. This need only mean that a number of lineages of flagellates, some more primitive than others, had originated before the diversification of the extant host groups. Presumably one would detect the operation of Szidat's rule *within* any such lineage.

Coevolution and historical biogeography

Systematists have traditionally been more interested in the utility of parasites for resolving problems such as host phylogeny and biogeographical history than historical ecology per se, and have raised the possibility that biogeographical and host parasite studies might be used in conjunction with one another (see Manter's Rules) to "reciprocally illuminate" issues of both kinds (see Hennig, 1966). A typical case from the early literature involves the "southern frogs" (family Leptodactylidae). These were regarded by some as necessarily polyphyletic because their disjunct distribution across the southern hemisphere was in conflict with the widespread theory that animals found presently in the southern hemisphere migrated there from the north under duress of competition (see Nelson and Platnick, 1981). Metcalf (1929, pp. 3-4) argued as follows in response:

In the recta of Australian and American southern frogs occurs a characteristic ciliate protozoan, *Zelleriella*, one of the Opalinidae, and some of these Australian and South American ciliates are almost if not quite specifically identical. This genus of ciliates is absent from the old world (except Australia) and in the New World is southern.... The parasites ... indicate seemingly beyond question that the Australian and American southern frogs are related and also that they arose in the Southern Hemisphere and passed by some southern route from one to the other of their southern habitats. It might be possible, however unlikely, that the southern frogs of Australia evolved from very ancient ancestors in a way parallel to that of the South American frogs, though almost always in cases of parallel evolution there are found some genetic criteria to distinguish such resemblance from that due to genetic relationship. But no one can for a moment believe that, along with the parallel evolutions of the American and Australian hosts, there was also a parallel evolution of their opalinids.

Metcalf used the congruence of two independently derived hypotheses of monophyly as evidence favoring both and argued further that the "linkage" of the same disjunct land areas by two independent sister groups of animals suggested a similarly close historical relationship between these areas. Since the publication of his papers, the evidence of plate tectonics has shown that these areas were, in fact, in contact.

From the point of view of the present chapter, the potential utility of broadening phylogenetic coevolutionary studies to include biogeography is that we might thereby increase our ability to distinguish between hypotheses of colonization and of association by descent. Consider a case involving only a few "host disjunctions," each fitting equally well both host phylogeny and some plausible theory of host transfer. For example, the African ostrich and the South American rhea share parasite taxa (species or genera) of several kinds, including lice, roundworms, and tapeworms (see Hennig, 1966, p. 179; Eichler, 1948). One might suppose that, were the hosts not geographically isolated, parasites should transfer with relative ease between them.

However, it is highly unlikely that the parasites of one or the other could have crossed the south Atlantic in the absence of a host. If a parasite phylogeny is compatible with both a plausible host transfer sequence and a host phylogeny, additional concordance of the latter with a geologic sequence may resolve the issue in its favor. In the case of the parasites of the ostrich and the rhea, mutual descent seems likely, because these birds are thought to have diverged when Africa and South America became separated by continental drift.

In closing this section we will cite one last case that serves to illustrate the great variety of animal associations challenging the ecological phylogeneticist. Among the numerous organisms found associated with termite nests are 100-odd genera of staphylinid beetles, representing approximately 11 independent invasions of this habitat (see review by Kistner, 1969). The ecology of the interaction is poorly known, but there is indirect evidence of marked adaptations of the beetles to life in the nest. For example, many forms exhibit a remarkable convergent enlargement of the abdomen, known as physogastry, whose utility is unknown. A number of other species are marked by a rounded, flattened "limuloid" body shape. Various authors have observed mutual grooming or licking between staphylinids and termites, and many of the beetles have become strongly dependent on the microclimate and food sources provided by the nest. The relationship is best regarded as commensal, rather than mutualistic, because the termitophiles are relatively rare and many nests lack them entirely.

Most termitophilous staphylinids are strongly host specific, and their distributions generally parallel those of their hosts, even when the latter are spread across several continents. These facts suggest that individual beetle lineages, once having taken up this way of life, might have subsequently undergone associated descent with their hosts. Seevers (1957) examined this issue in detail and concluded that most evolution in the termitophiles had indeed occurred in association with that of their hosts. Two of the phylogenies depicted by Seevers are presented in Figure 8. If we consider just the hosts occupied by the subtribe Corotocina (Figure 8A), there is some support for association by descent, in that of the three possible arrangements of the host genera, the one best fitting the beetle phylogeny corresponds to termite relationships as presently understood. However, there remains the unresolved problem of the large number of phylogenetically intervening but unoccupied host groups. Although some of these absences may simply reflect limited sampling, it would seem important to search for geographical or ecological factors predicting the absence of termitophiles from particular host lineages. The Termitogastrina, on the other hand (Figure 8B), show no strong evidence of mutual descent. The termitophilous staphylinids present just

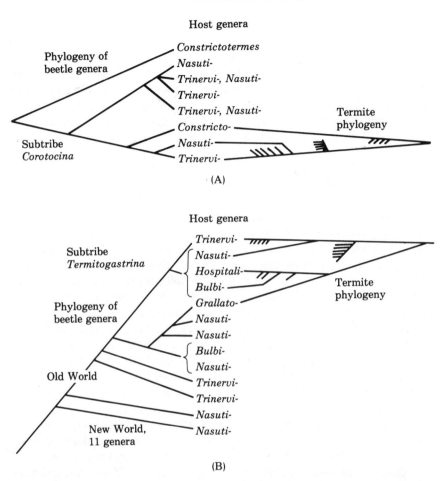

FIGURE 8. Phylogenies of two groups of staphylinid beetles obligately associated with termite nests, following Seevers (1957), together with host termite phylogenies according to Krishna (1970). Small, unlabeled branches on termite cladograms represent genera not known to harbor members of beetle group. All termite names end in *-termes*.

enough suggestion of a coevolving fauna to make further investigation highly desirable. We must note, finally, that for each case discussed in this section, associated descent may not be evident overall, yet could occur *within* the lowest-level taxa being considered, that is, on a shorter time scale than that involved in the establishment of host association differences among these taxa. Over very short time spans, *most* associations are undoubtedly "inherited." The question to be asked is how long such evolutionary continuity of association typically persists.

Phytophagous insects

Much of the current surge of interest in coevolution can be traced to a paper about butterflies and their host plants (Ehrlich and Raven, 1964). It therefore seems appropriate to review some of the evidence regarding the degree to which phytophagous insect groups and their hosts have evolved in association. The model implicit in Ehrlich and Raven's discussion and developed explicitly by Benson et al. (1975) involves not strict, that is, continuous, mutual descent of hosts and herbivores, but alternating periods of diversification of each. A hypothetical phylogenetic pattern of the sort that might be expected under their model is shown in Figure 9. The two long branches in the host phylogeny represent lineages that have developed some novel form of immunity to attack by herbivores; this development allowed them to undergo a period of relatively rapid diversification. Each long branch of the herbivore tree may be taken to represent a line that has evolved some method of overcoming the defenses of a hitherto-immune plant group, a method that enabled it in turn to radiate across those already-diverse hosts. Each pairwise association is a result most immediately of colonization, but host phylogeny will still be a significant, albeit imperfect, predictor of the evolution of host associations. Even though there is no correspondence of phylogenies within "radiations," earlier-originating herbivore groups will tend on the average to be found on earlier-originating plant groups, if only because these were the sole hosts present when those herbivores were differentiating. The higher-level pattern of mutual descent will be most apparent when the colonizers of each recently diversified host taxon are derived from forms previously occupying the nearest relatives, and hence some not-too-distant ancestor, of that host group (Figure 9). If no such stricture applies, the pattern may be all but completely obscured, meaning that the points of historical contact between any two associated lineages (i.e., the periods during which respective ancestors giving rise to them were associated) have been few. Then the opportunity for mutual historical influence between particular lineages would have been limited, even though the evolution of hosts and herbivores might have been broadly contemporaneous. Each lineage may in this case be regarded as adapting to a background composed of numerous actual or potential associates, a process that has been termed "diffuse" coevolution.

How might one distinguish between the "reciprocal radiation" model of Ehrlich and Raven and a strict hypothesis of either colonization or mutual descent? Neither the methodology nor the evidence required are yet available, but a brief inquiry into this issue may il-

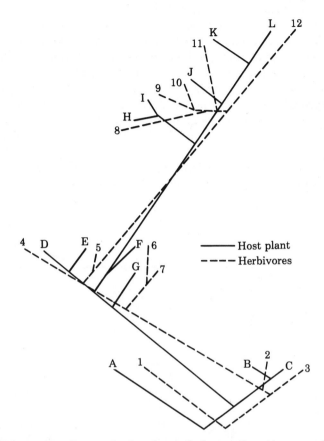

FIGURE 9. Artificial example showing a phylogenetic pattern possible under "reciprocal adaptive radiation" model of coevolution. For explanation, see text.

lustrate some of the difficulties involved in deciding among similar historical theories. The "reciprocal radiation" model leads to several expectations. First, there should be a nonrandom fit overall to association by descent. Second, the deviations from this fit should themselves be predictable, occurring within but not among groups of hosts that differ in antiherbivore innovations that can be recognized on independent grounds. Thus, the best single predictor of the evolution of host preference should be the ordering of "defensive novelties" specified by the host phylogeny. The expectation so established, however, might also be compatible with a simple host-shift model (the "sequential evolution" of Jermy, 1976b), under which each host is colonized from

85

the host with the most similar "defense" characteristics. The more nearly the evolution of defenses is strictly divergent, the less distinguishable these alternatives will be.

The two theories may be separable, however, if unrelated hosts frequently have converged in their defenses, as have rutaceous and umbelliferous plants that produce the same essential oils. The strict host-shift theory predicts a close match to similarities in defenses, because all hosts are available simultaneously for colonization. Under the "reciprocal radiation" model, by contrast, host transfers should be more closely tied to phylogeny, because the convergent host may be unavailable (e.g., not yet evolved) when a particular transfer is occurring.

Although Ehrlich and Raven (1964) held that "plants and phytophagous insects have evolved in part in response to one another . . .

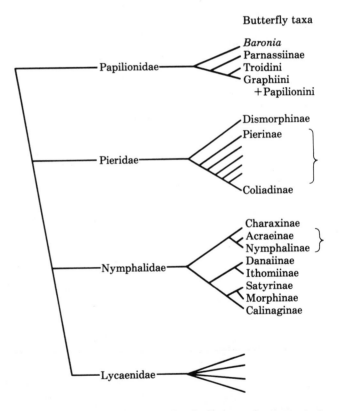

FIGURE 10. Summary of recent ideas on butterfly phylogeny. Arrangement follows Ehrlich (1958), except as modified by Kristensen (1975) and as follows: phylogeny of Pieridae after Klots (1931); arrangement of Papilionidae after Munroe and Ehrlich (1960). Hosts listed are major ones according to Ehrlich and Raven (1964). Slash (/) separating hosts indicates close relationship. Ar-

86

in a stepwise manner," they drew relatively few conclusions about the temporal order of "radiation" of particular groups. Their major concern was to establish that many higher taxa of butterflies are largely restricted to sets of hosts sharing a property (namely, secondary chemistry) known to strongly influence ecological interactions. Many of the hosts so grouped are phylogenetically distant, suggesting that host transfer has been a major mode of establishment of new butterfly habits. The possibility that there has nevertheless been broad-scale coevolution between butterflies and their food plants is not strongly supported by the sketchy outlines of butterfly and angiosperm phylogenesis currently available. To be sure, one might argue from Figures 10

Major hosts

Fabaceae
Aristolochiaceae
Aristolochiaceae
"Woody Ranales" → Rutaceae, Umbelliferae,
  many others
  Rosaceae
Fabaceae

Cruciferae/Capparaceae →

Fabaceae and others

"Woody Ranales" and many others
Violales/Passiflorales; Urticales/
  Euphorbiales; and many others
Asclepiadaceae/Apocynaceae
Asclepiadaceae/Apocynaceae → Solanaceae
Grasses and other monocots
Variety of both dicots and monocots
Urticales

(Very large and poorly known group; great
diversity of feeding habits, large-scale
patterns rarely evident; many associated
with ants)

rows indicate probable secondary feeding habits. Ithomiinae feed mostly on Solanaceae, but earliest ones probably ate Apocynaceae (see Edgar et al., 1974; Gilbert and Ehrlich, 1970). "Woody Ranales" include families in both Magnoliales and Laurales (see Figure 11).

and 11 that, on average, the earliest-originating lines within major butterfly taxa feed on relatively ancient hosts. Thus, for example, the aristolochiaceous and "woody ranalian" hosts fed on by the main line of papilionids are older (or have more primitive secondary compounds) than either the rutaceous, umbelliferous, or rosaceous plants (among others) representing more derived habits within this family, or the dilleniid plants that are widely fed on by more recent groups such as Pieridae and Nymphalidae. However, major exceptions to this generalization are apparent. For example, the earliest diverging lineages of both Papilionidae and Pieridae feed on legumes (Fabaceae), one of the most recent and highly evolved of dicot families. The fact that many more ancient lepidopteran groups (see below and Figure 12) are associated with advanced, as opposed to primitive, dicots also suggests that the ancestral butterflies fed on advanced hosts. Much additional work on both papilionoid and angiosperm relationships, particularly among the Nymphalidae and their dilleniid hosts, will be required to fully settle the issue, but the evidence to date argues against a significant role for mutual descent in establishing the host preferences of the larger groups of butterflies (see also Vane-Wright, 1978).

Even if the Papilionoidea as a whole invaded their host taxa largely after the major taxa of plants had evolved, "reciprocal radiation" might nevertheless have operated over much longer or shorter time scales than those of the divergence of butterfly families. Perhaps, for example, the diversification of angiosperms, with subsequent colonization by butterflies, represents just one episode in a large-scale coevolutionary association of land plants and the order Lepidoptera as a whole. In the much-condensed view of lepidopteran relationships depicted in Figure 12, the first two host disjunctions are compatible with such a notion, although one would need to account for the absence of any ancient moth lineages from both ferns and monocots (on both of which several advanced groups of Lepidoptera feed), as well as the considerable disparity in the estimated ages of the two groups. Even if the very earliest lepidopteran and land plant ancestors evolved in concert, however, there is little indication that the main line of evolution giving rise to the major groups of extant Lepidoptera was associated with successive plant radiations. Some of the oldest angiosperm-inhabiting moth groups (e.g., the Eriocraniidae) are associated primarily with relatively young host groups, whereas primitive hosts characterize only a few, relatively recent, higher lepidopteran taxa (e.g., Papilionidae). A similar pattern was described earlier for the major groups of termite-inhabiting flagellates, however, and there remains the possibility that diversification of early lepidopterans on early angiosperms gave rise to a multitude of lineages (e.g., butterflies), each of which subsequently diversified in tandem with angiosperm phylogenesis

Plant taxon

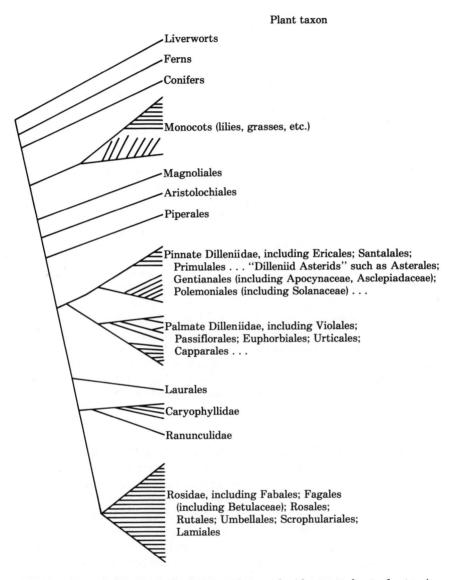

FIGURE 11. An informal phylogeny of some lepidopteran host plants. Arrangement of angiosperms (monocots and above) follows Hickey and Wolfe (1975). Numbers of branches within larger taxa indicate approximate rank of number of orders included. Recent evidence will necessitate some changes in this scheme (L. J. Hickey, personal communication), and other workers would give somewhat different arrangements.

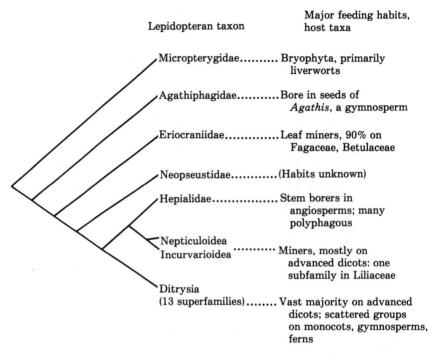

| | Major feeding habits, |
| Lepidopteran taxon | host taxa |

Micropterygidae.......... Bryophyta, primarily
                         liverworts

Agathiphagidae........... Bore in seeds of
                         Agathis, a gymnosperm

Eriocraniidae.............. Leaf miners, 90% on
                         Fagaceae, Betulaceae

Neopseustidae............ (Habits unknown)

Hepialidae................. Stem borers in
                         angiosperms; many
                         polyphagous

Nepticuloidea
Incurvarioidea ·········· Miners, mostly on
                         advanced dicots: one
                         subfamily in Liliaceae

Ditrysia
(13 superfamilies)........ Vast majority on advanced
                         dicots; scattered groups
                         on monocots, gymnosperms,
                         ferns

FIGURE 12. Early evolutionary history of Lepidoptera, following Kristensen and Nielsen (1980). Food plant information from Powell (1980). Ditrysia contain vast majority of living species. "Advanced dicots" means (largely) Dilleniidae and Rosidae (see Figure 11). (Note added in proof: More recent information suggests that Micropterygidae will feed on a variety of plant material found in the litter layer; D. Davis, personal communication.)

(Powell, 1980). Much greater resolution of relationships within the groups depicted here will be required to test such a theory. Available fossil dates suggest roughly contemporaneous radiation of Lepidoptera and Angiospermae. Suppose, as seems likely, that future investigation fails to demonstrate mutual descent with angiosperms within each of the major lepidopteran lineages. This could mean that, whereas the oldest ancestors of the hosts and phytophages were themselves associated, host transfers have in the meantime been so pervasive that present-day associations have no significant joint history (see Powell, 1980). It could also mean that extinction has obliterated the pattern. (See caption to Figure 12 for information added in proof.)

As we have seen, many relatively low-level lepidopteran taxa (e.g. tribes and genera) are narrowly restricted to particular angiosperm groups. Even if these associations themselves were established primarily by host transfers, mutual descent may have been important in their subsequent elaboration. In very few instances, however, is there enough systematic information to examine this question.

Perhaps the best-studied case is that of the heliconiine nymphalid butterflies, a largely Neotropical group of approximately 70 species in 11 genera, all of which feed on Passifloraceae. This association was examined from a systematic and coevolutionary point of view by Benson et al. (1975), on whose discussion ours will rely heavily. A phylogeny and associated character information for the heliconians were presented by Brown (1981; see Figure 13); the somewhat different arrangements depicted by Brown (1972) and Emsley (1965) lead to the same conclusions to be drawn here. The hosts listed in Figure 13 are, exception as noted, those designated by Benson et al. (1975), as the primary ones for each butterfly taxon; many of the species have been also recorded from other passionflower groups. Benson et al. arranged the genera of new-world passiflors according to presumed degree of "evolutionary advancement" with respect to a number of floral and vegetative characters (following Killip, 1938). Although recognizing, as did these authors, the danger of equating degree of advancement with cladistic sequence, we can use this assumption to obtain a rough estimate of the prevalence of mutual descent in this assemblage. The evidence at hand appears to support the hypothesis of coevolution only weakly: the first and second divergences, isolating groups on the relatively primitive *Adenia* and *Passiflora* section *Astrophea*, agree fairly well with this explanation. However, on the rest of the heliconian phylogeny, primitive and advanced hosts seem to follow each other in no discernable order, and many quite different transformation series for the host association "character" would fit the observations equally well. Although a more rigorous analysis is clearly desirable, one cannot rule out the possibility that most (perhaps all) of the evolution of heliconians occurred after the diversification of the Passifloraceae, including that of *Passiflora* into its various sections.

Benson et al. (1975), whose largely implicit analysis seems to have treated the heliconian phylogeny not as a whole but as a series of independent radiations and regarded the "primitive" forms of each as indicating the ancestral feeding habits despite their sometimes nonbasal cladistic positions, reached the opposite conclusion. The "reciprocal radiation" theory would gain credence if just the apparent exceptions to it were especially likely to represent host transfers. Thus, Benson et al. (1975) argued that the *sara-sappho* Heliconius were able to "reradiate" onto the primitive *Passiflora* section *Astrophea* because they used a different part of the plant (meristems versus mature leaves) from earlier heliconians using the same hosts. Work in progress on

91

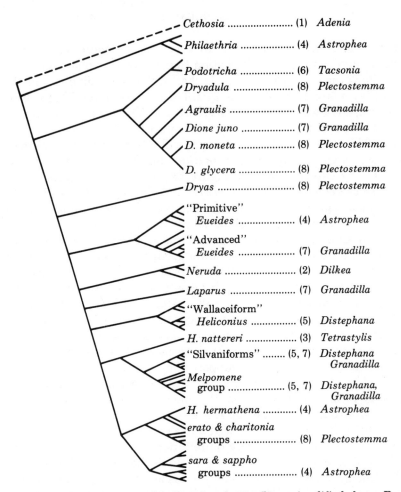

Heliconiines		Major host taxa

Cethosia	(1)	Adenia
Philaethria	(4)	Astrophea
Podotricha	(6)	Tacsonia
Dryadula	(8)	Plectostemma
Agraulis	(7)	Granadilla
Dione juno	(7)	Granadilla
D. moneta	(8)	Plectostemma
D. glycera	(8)	Plectostemma
Dryas	(8)	Plectostemma
"Primitive" Eueides	(4)	Astrophea
"Advanced" Eueides	(7)	Granadilla
Neruda	(2)	Dilkea
Laparus	(7)	Granadilla
"Wallaceiform" Heliconius	(5)	Distephana
H. nattereri	(3)	Tetrastylis
"Silvaniforms"	(5, 7)	Distephana Granadilla
Melpomene group	(5, 7)	Distephana, Granadilla
H. hermathena	(4)	Astrophea
erato & charitonia groups	(8)	Plectostemma
sara & sappho groups	(4)	Astrophea

FIGURE 13. Phylogeny of heliconiine butterflies, simplified from Brown (1981). Names in quotes after Benson et al. (1975). Hosts listed are "preferred plant taxa that are assumed to have been responsible for past radiation" (Benson et al., 1975) of each butterfly group, except that host for *Podotricha* is from Brown (1981) and for *Cethosia* from Corbet and Pendlebury (1978). Numbers in parentheses indicate rank in "degree of advancement," and hence approximate age, of host groups, according to Benson et al. (1975). (*Adenia*, not treated by Benson et al., placed by their criteria.) Ranks 4–8 are sections of *Passiflora*; all others separate genera.

both the systematics and ecology of this system (see, for example, review in Brown, 1981) may eventually resolve these problems. The heliconians and their relatives might be regarded as a test case for mutual descent: the dominant hosts for many of these butterflies are

the families Passifloraceae, Violaceae, Turneraceae, and Flacour-
tiaceae, an assemblage coherent on phylogenetic grounds but for
which Ehrlich and Raven (1964) could identify no secondary chemical
similarity, though they predicted that one would be found.

Other phytophagous groups can be found for which mutual descent
also seems plausible. Thus, Chemsak (1963) postulated parallel evolutionary development of the longhorn beetle genus *Tetraopes* with its
milkweed food plants (*Asclepias*), although he did not present a
phylogenetic analysis. An especially intriguing candidate is the association of the hundreds of species of figs and their highly specific wasp
pollinators, family Agaonidae (see Wiebes, 1979; Chapter 13 by Feinsinger). Ramirez (1975) presented cladograms showing complete agreement with mutual descent for a group of agaonid genera and their host
figs. Until a number of such cases have been examined phylogenetically, it is not possible to state whether mutual descent, including
"reciprocal radiation," plays a significant role in the evolution of herbivorous insects and their hosts at this taxonomic level. It will also be
of great interest to determine whether mutual descent operates at the
lowest level accessible to phylogenetics, that is, the differentiation between races or species, as postulated for cynipid oak gall wasps by Cornell and Washburn (1979). The species in the two genera that have
been studied in most detail, *Cynips* and *Neuroterus* (Kinsey, 1923,
1930, cited in Cornell and Washburn, 1979), exhibit large numbers of
"varieties," the most closely related of which generally occur on most
closely related oaks, which tend to be allopatric. This would appear to
be a test case for species-level associated descent, and further investigation is greatly to be desired.

THE STUDY OF COLONIZATION

We chose for the preceding examination several cases in which mutual
descent seemed most likely. In many other groups of phytophagous insects, however, the taxonomic separation between hosts of even sister
phytophage species allows one to reasonably discount mutual descent
out of hand (see Chapter 10 by Futuyma for examples). Indeed, the
evidence available suggests that host transfer is by far the predominant mode of establishment of new host associations among insect
phytophages, as it is for many other kinds of association. For this
reason, it is appropriate to consider in more detail what generalizations might be made about colonizations.

Of the many questions that can be asked about host transfer, the
most germane is that of which host characteristics explain the evolutionary sequences of host associations inferred from symbiont

cladograms (see Mickevich, 1981). Among the many factors that might affect the likelihood of transfer between pairs (or sets) of hosts, we will distinguish between those that affect the likelihood of encounter of a given host by residents of another and those that determine whether such encounters will result in the evolution of a new association.

The probability of encounter, including such factors as physical proximity of different hosts, has been proposed as a dominant influence in the evolution of host associations for a variety of symbionts. Many phytophagous insects, for example, are closely bound to particular types of habitat, often more so than their hosts. In such cases the diets of oligophagous or closely related monophagous herbivores frequently include plant species whose major commonality appears to be their habitat association. Many examples could be cited; Hering (1954) discussed the case of the moth *Euspilapteryx phasianipennella*, which feeds, in moist habitats, on the related families Chenopodiaceae and Polygonaceae, but also on *Lysimachia* (Primulaceae) and *Lythrum* (Lythraceae), which are hosts that are taxonomically distant from the former pair but are characteristic of wet places. Similarly, the shift of an originally rodent-infesting group of fleas onto burrowing owls (Rothschild and Clay, 1952, cited in Ross, 1962) is readily explained by the hopping of parasites from prey to predator.

Intrinsic differences in hosts' ability to attract and support the growth and reproduction of particular symbionts are also likely to determine host transfer sequences, especially in very vagile groups for which different hosts are readily accessible. Several examples were cited earlier, including the apparent importance of quill size in the colonization of bird species by feather mites (Kethley and Johnston, 1975) and of plant secondary chemistry in the evolution of phytophagous insect diets.

Although a number of workers have assessed in some fashion the importance of one or more effects in the evolution of a particular set of associations, there has been almost no attempt at fitting models combining factors of different types to host transfer sequences inferred from symbiont cladograms. In both methodological and empirical respects, the phylogenetic study of host shift sequences is in its infancy. Thus, Ehrlich and Raven (1964) argued the importance of plant chemistry in butterfly evolution on the basis of broad associations of butterfly taxa with sets of chemically similar plants; in almost no cases do they describe the kind of multistep sequence needed to determine the degree to which invasion of new hosts is truly "stepwise" with respect to host chemistry. (A possible example of such "stepwise" colonization is the evolution of the umbellifer-feeding habit in some papilionids from an ancestor originally on the chemically dissimilar "woody Ranales"; the "gap" may have been bridged by a transfer first to Rutaceae, some of which share alkaloids with "woody Ranales" but also attractant compounds with the umbellifers; Ehrlich and Raven, 1964.) The phylogenetic study of colonization "rules" presents many of the same problems encountered in our discussion of mutual descent. In particular, there are likely to be so many causes operating simultaneously that sorting out their contributions within any one group may be very difficult. A "faunistic" approach, as in our earlier example of the parasite community on crocodilians, might be used to advantage here as well.

We have been concerned above with the question of constraints on the evolutionary "accessibility" of hosts from the point of view of symbionts, in much the way that systematists ask whether particular states of a morphological character must lie on a historical path between certain others. It is possible for such constraints to exist and yet not obviously correspond to any generalization about host properties. Thus, Ehrlich and Raven (1964) identified a number of butterfly host shifts that did not make obvious "sense" on chemical grounds. These "exceptions" may be explainable by some other as-yet-unstudied general factor, but they might also be due to highly idiosyncratic preadaptation (see Mitter and Futuyma, 1983). It is also possible that, at least within a restricted set, all hosts may be more or less equally colonizable from one another, host choice evolution being guided only by localized and temporally varying ecological factors. Thus, Gilbert and Smiley (1978) argued that the potential food plant range (as judged by larval fitness) of many *Heliconius* species is much greater than that expressed in nature and suggested that geographical variation in host preference in these butterflies may be due to "ecological monophagy," that is, localized selection pressure on female oviposition choice imposed by competition, parasitism, etc. (see also Fox and Morrow, 1981). When host transfer sequences appear to follow no generalization, experimental study (on preadaptation or on the ecological circumstances surrounding recent host transfers) may help to decide between the alternatives of "no intrinsic constraints" and "idiosyncratic constraints" on colonization histories.

Many other questions can be asked about evolutionary host transfer. For example, there has been much interest in the evolution of varying degrees of feeding specialization among phytophagous insects. An oft-repeated suggestion in the earlier literature (see, for example, Brues, 1920) was that, since many "primitive" herbivores, namely, grasshoppers and their relatives, tend to have relatively broad diets, there may have been an overall trend for the evolution of monophagy from polyphagous beginnings, perhaps because of the increasing diversity of plant defenses. It is now recognized, however, that the phytophagous habit has been acquired independently by a number of major insect groups, for example, the Lepidoptera. Within this group, the observed trend is, if anything, opposite to Brues' suggestion (Figure 13). The oldest lineages that feed on seed plants (e.g., Agathiphagidae, Nepticuloidae, Eriocraniidae) are mostly composed of mono- or oligophages, and polyphagy appears to be a derived condition that has developed independently in a number of primitively specialized-feeding lineages. Closer examination might reveal instances of the reverse trend, and very little is known about smaller-scale patterns in the evolution of this trait.

Distinct from the problem of what *sequence* the evolution of association will follow, there are the issues of how often and under what circumstances host transfers should occur. Although phylogenetic evidence clearly bears on the latter as well, limitations of space prevent its consideration here (but see, for example, Bush, 1975; Gilbert, 1979; Mitter and Futuyma, 1983; Ross, 1972). The two questions are not wholly separate; for example, one would like to know how and why the rules governing host shift sequences vary among lineages (see, for example, Gilbert, 1979).

EVOLUTION IN ASSOCIATED LINEAGES

We have thus far been concerned mainly with how and when ecological associations originate. A phylogenetic approach can also aid in the study of the evolutionary consequences of such association. Our discussion of this topic must be brief and largely hypothetical because of the rarity of studies providing the necessary combination of systematic and ecological information.

Separating "common causes" from coevolution

The fundamental postulate of coevolution is that interacting species influence each other's evolution. This hypothesis can be tested by systematics by finding some form of temporal relatedness of evolutionary events (character changes or divergences) between associated lineages. However, the evolution of a set of such lineages could be governed entirely by the independent responses of each to a shared history of geographical isolating events, producing matching temporal patterns of speciation and perhaps even of rates of character change, with no requirement for "host tracking" or any other form of coevolution *sensu stricto* (see Brooks, 1979b). Thus it is necessary to distinguish "correlated evolution" due to coevolution (the "coaccommodation'" of Brooks, 1979b) from that due to noninteractive causes.

Phylogenetic studies might contribute in several ways to the evidence for adaptation caused by coevolution (see Chapter 1 by

Futuyma and Slatkin). Experiments designed to demonstrate a fitness differential capable of accounting for the evolution of a characteristic compare that feature with some alternative type, often constructed artificially (see, for example, Bentley, 1976). The use of phylogenetic evidence to select a form closely resembling the ancestral morph from which the putative adaptation evolved may allow a better description of the actual historical change. For example, Benson (1972) performed an elegant test of Müllerian mimicry between the butterflies *Heliconius erato* and *H. melpomene* in Costa Rica, by comparing the fitness of control *erato* to that of specimens modified to look like the race endemic to a region of Colombia (see also Chapter 12 by Gilbert). Suppose, however, that we wished to determine experimentally whether the selection for Müllerian mimicry could have been discriminating enough to account for the evolution of the Costa Rican morph from its immediate ancestor. Then the most appropriate "experimental" form should be one constructed to resemble as closely as possible the phylogenetically inferred ancestral type [see Turner (1981) for phylogenies of some of the morphs of these two species].

Although it may often be difficult to resolve a particular question of adaptation considered in isolation, it may nevertheless be possible to show that the case is part of a clear-cut larger pattern, involving analogous features evolved independently under similar circumstances. Such a finding would be strong evidence that the particular analog was an adaptation of some sort. In other words, convergence, which phylogenetic methods are designed to detect, can be a strong source of evidence on adaptation. An illustration is provided by the phenomenon of physogastry in termitophilous insects, which occurs independently as a derived character in a number of unrelated lineages (see Kistner, 1969) and is not present in earlier-originating termitophilous forms. From these observations we can rule out the possibility that physogastry is a "preadaptation" that allowed its possessors to invade termite nests or that it reflects a purely environmental influence of termite nests on termitophile development. (For an argument of similar kind, see Chew and Robbins, 1982.) Therefore we can have considerable confidence that physogastry evolved as an adaptation to termitophily, even though its actual function is poorly understood.

There does not seem to be any criterion for judging hypotheses of "reciprocal coevolution" apart from those used to test the putative individual adaptations they entail. Moreover, tests of adaptation may be made difficult in coevolving systems by the extinction of intermediate stages. However, in a sufficiently elaborately coevolved system in which intermediates are extant, it may be possible to apply the criteria of temporal ordering and repeatability of change to make inferences

97

about the course of adaptation. To take a hypothetical example, suppose we wish to show that in a host–parasite system (1) some hosts have features that evolved because of parasite pressure and (2) some parasites have traits that evolved because they permitted circumvention of evolved defenses. Phylogenetic evidence for the first of these would comprise the appearance of some type of feature (e.g., a new secondary compound in the case of plants) preferentially in lineages primitively exhibiting a high level of parasite attack. Evidence for the second would comprise preferential occurrence of novel states of some host-related character in parasites feeding on hosts bearing evolved defenses, as opposed to their relatives exhibiting more primitive habits. We are not aware of any case presenting all of these features, but some apparent instances of "stepwise" evolution of secondary compounds in plants attended by successive restriction of herbivore faunas [e.g., the appearance of cucurbitacins in some Cruciferae, normally characterized only by mustard oils (Chew and Rodman, 1979)] merit further study. A number of authors have speculated on the long-term course that such reciprocal adaptation should take (see, for example, Futuyma, 1979, and Chapter 10; Wilson, 1980).

Associates evolving together may affect each others' rates of diversification, in addition to evoking reciprocal adaptations. For example, one implication of the model envisioned by Ehrlich and Raven (1964) is that the temporary escape from parasite attack conferred by a novel defense should allow the host lineage bearing it to become more diverse than its sister group (see Figure 1). These authors speculated that the enormous diversity of angiosperms, as compared to other seed plants, might have resulted from the acquisition of the secondary compounds now found in their primitive representatives. A similar trend should hold for parasite groups colonizing new host taxa. Such seemingly straightforward phylogenetic predictions have yet to be quantitatively evaluated for any parasite or host group.

CONCLUDING REMARKS

In this essay we have attempted to identify some major questions to be addressed by the phylogenetic study of ecological associations, to outline some of the logic useful in answering them, and to sample the range of ecological–phylogenetic patterns found in nature. It is clear that both the evidence and the methods required for such inquiries are in a primitive stage of development. It is thus too early to tell either how much regularity ecological history exhibits or how much resolution its phylogenetic study will permit. It is our conviction, however, that the combination of ecological and systematic approaches pioneered by Ehrlich and Raven (1964) may ultimately lead to a sophisticated understanding of community evolution.

98

CHAPTER 5

COEVOLUTION
IN BACTERIA
AND THEIR VIRUSES
AND PLASMIDS

Bruce R. Levin and Richard E. Lenski

INTRODUCTION

Populations of bacteria can be parasitized by a variety of independently replicating genetic molecules. Based on functional rather than phylogenetic considerations, these replicons are classified as *plasmids* or *viruses*. Plasmids are extrachromosomal molecules of DNA that are present in one or more copies per cell. They replicate at the same average rate as the host chromosome and, in the course of cell division, are transmitted to descendants of infected cells with high frequency. In addition to this capacity for vertical transmission, some plasmids have specific adaptations for horizontal (i.e., infectious) transmission. By processes that require cell–cell contact, copies of these *conjugative* plasmids may be transmitted from *donor* cells to *recipient* cells. For reviews of the basic biology of plasmids, see Meynell (1973), Falkow (1975), and Broda (1979).

Bacterial viruses (*bacteriophage*, commonly contracted to *phage*) differ from plasmids in that they can exist outside the cell, encapsulated in a protein coat that both augments their extracellular term of survival and enables them to attack sensitive cells. Phage infection commences with the adsorption of the virus to specific receptor sites on the bacterium and the passage of its genetic material (DNA or RNA) into that cell. For *virulent* bacteriophage, replication is necessarily by a *lytic* cycle, which terminates with the death of the

99

host and the release of large numbers of phage particles. *Temperate* bacteriophage may also go through this lytic replication cycle, but there is a certain probability that following adsorption the viral DNA will be maintained as a *prophage* that is stably inherited by the descendants of the originally infected cell. This prophage may be integrated into the host chromosome or, in the case of some temperate bacteriophage, exist as a plasmid-like extrachromosomal element. At rates that depend on environmental conditions, individual bacteria carrying prophage—*lysogens*—will be *induced* and go through a lytic cycle that terminates with the death of the host cell and the release of phage particles. For reviews of the basic biology of bacterial viruses, see Adams (1959), Stent (1963), and Stent and Calendar (1979).

Bacterial viruses and plasmids are parasites in the sense that they have no host-free mode of reproduction and do not unconditionally increase the likelihood of their hosts surviving and reproducing (a definition of parasite somewhat broader than that in Chapter 9 by May and Anderson). Indeed, in the case of virulent bacteriophage, the cost of infection for individual cells is quite dear, and even the carriage of seemingly innocuous plasmids may impose a burden on bacteria. On the other hand, for many plasmids and some phage, the association between these autonomous replicons and their hosts is more that of mutualists. Resistance to antibiotics and heavy metals; the capacity to produce restriction enzymes, toxins, bacteriocins, and antibiotics; the ability to ferment certain carbon sources; and the production of structures for the invasion of specific habitats are characteristics that are often determined by plasmid-borne, rather than chromosomal, genes. (For reviews of the various kinds of plasmid-determined bacterial phenotypes, see Novick, 1974; Chakrabarty, 1976; Broda, 1979; and Davey and Reanney, 1980.) Although the abundance and diversity of phage-determined bacterial phenotypes is less than that of plasmid-determined phenotypes, there are some cases where antibiotic resistance and toxin production are due to genes borne on the prophage of temperate viruses.

Conjugative plasmids and temperate and sometimes virulent bacteriophage also play a role in bacterial adaptation and evolution by serving as vehicles for the exchange of genetic material. In the course of their infectious transmission, these autonomous replicons may pick up chromosomal genes or non-self-transmissible plasmids from one bacterium and may transmit them to another. Because the host ranges of bacterial plasmids and viruses often exceed "species" bounds and because there are mechanisms for recombination in the absence of close genetic homology (insertion sequences and transposable genetic elements; see Calos and Miller, 1980), the range of gene exchange mediated by plasmids and phage can encompass very phylogenetically diverse groups of bacteria.

It seems clear that the association between bacteria and their viruses and plasmids is a very ancient one. Phage and/or plasmids have been found in virtually every species of bacterium that has been examined for their presence. More than 90% of genetically distinct clones of *Escherichia coli* carry at least one plasmid, and the majority of these carry more than one (see, for example, Caugant et al., 1981). It has been estimated that nearly 100% of naturally occurring members of the genus *Pseudomonas* are lysogenic for some temperate virus (Holloway, 1979). For *E. coli* and closely related Enterobacteriacea, more than 200 plasmids and more than 80 species of phage have already been described (Novick, 1974; Reanney, 1976).

Plasmids and phage (as well as their bacterial hosts) can accumulate genetic variability through mutation, recombination, and the acquisition or loss of transposable elements. Because plasmids and phage cannot reproduce outside their bacterial hosts and because these replicons influence their hosts' survival and reproduction, coevolution must be very significant to the overall evolution of bacteria and their plasmids and phage. In this chapter, we will consider the nature and consequences of this coevolution. Using simple models tailored to the specifics of the interactions (Slatkin and Maynard Smith, 1979; Chapter 3 by Roughgarden) between these replicons and their hosts, we will predict the direction of selection on the parameters governing the systems. We will compare these a priori considerations with empirical results obtained from experimental and natural populations of bacteria and their plasmids and phage.

COEVOLUTION IN VIRULENT PHAGE AND THEIR HOSTS: A PRIORI CONSIDERATIONS

A model

A schematic representation of the association between populations of virulent phage and host bacteria is presented in Figure 1. The mathematical model, also presented in the figure, is a modified version of one employed by Levin, Stewart, and Chao (1977) and is analogous to that developed by Campbell (1961). The model assumes that bacteria and phage are thoroughly mixed in a liquid habitat of constant volume; bacterial resources enter and populations are washed out at a constant rate ϱ.

In the absence of phage infection, the bacteria multiply (via binary fission) at a rate ψ, the intrinsic rate of increase of the bacterium under specified environmental conditions. Environmental conditions critical to bacterial growth include temperature and resource concentrations

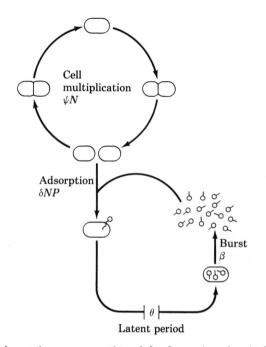

Adsorption
δNP

Burst
β

θ

Latent period

FIGURE 1. Schematic representation of the dynamics of a virulent phage and its bacterial host. N, Density of uninfected host cells; P, density of free phage; ψ, rate of cell multiplication; δ, phage adsorption rate parameter; θ, time between adsorption and burst of infected cells; β, number of phage produced per infected cell; ϱ (not shown in figure), rate of flow through the habitat and concomitant dilution of cell and phage populations. This model can be expressed as time-delay differential equations:

$$dN/dt = \psi N - \delta NP - \varrho N$$
$$dP/dt = \beta e^{-\varrho \theta} \delta N' P' - \delta NP - \varrho P$$

where N' and P' refer to the densities of cells and phage θ time units before the present.

(see Monod, 1949). Phage adsorb to sites on bacterial membranes at a rate that is proportional to cell density N and to the adsorption rate parameter δ. Infected bacteria are removed from the growing cell population because, as a result of infection, they cease multiplication and are fated to die. After a latent period of duration θ (during which the phage multiply in the host cell), a fated cell lyses and bursts, releasing β phage particles. Infected and uninfected cells and free phage are washed out from their populations at a rate ϱ that is independent of their densities.

Heritable variation in the parameters governing the growth and phage infection properties of the bacterial and phage populations can

102

result in differential rates of survival and reproduction, that is, natural selection. Therefore, by considering the effects of changes in the various parameters of the model, we can make inferences concerning the direction and intensity of selection in these populations.

Selection in the bacterial population

From the perspective of the host bacteria, selection should act to increase the rate of cell multiplication ψ, regardless of the density of phage. Because adsorption leads to cell death, selection acting on the cell population should reduce the rate of adsorption δ; the intensity of this selection will, however, be dependent on phage density. Thus, at low phage density, a mutation rendering a cell resistant to the phage may be disadvantageous if that mutation engenders a significant reduction in the cell's intrinsic rate of increase. The same mutation will likely be advantageous, however, when phage are abundant.

Because neither the burst size β nor the lag time θ directly enters the equation for bacterial growth in the mass-action model presented above, changes in these parameters should not be the direct result of selection acting on the bacterial population.

Selection in the phage population

The growth rate of the phage population is directly related to the burst size β and to the rate of adsorption to uninfected cells δ, and hence any increase in either of these parameters should be favored by selection acting on the phage.

Selection acting on the phage should reduce the latent period θ, although an examination of the time-delay differential equation for phage population growth does not immediately reveal this. An increase in the latent period effectively reduces the burst size, as a result of the washout of infected cells. Because "progeny" phage are subject to the same rate of washout ϱ, whether they are in infected cells or free, the advantage of reducing the latent period is *not* related to ϱ. Instead, the advantage of shortening the latent period lies in the earlier opportunity it provides for progeny phage to infect new host cells and further multiply.

Antagonistic selection and persistence

Unilateral selection in the bacterial population could result in the elimination of the phage from the habitat. If more resistant cells (those with lower δ) replace the more sensitive and if the phage are unable to increase when rare in a population of these more resistant cells, the phage would be eliminated. In cases where resistance is complete (i.e., $\delta = 0$), the resistant cells could completely displace the sensitive cells *only* if the former have an equal or greater rate of increase ψ. Thus, *if resistance engenders some cost in the competitive performance of the bacteria and if the sensitive cells are able to maintain a stable association with the phage, the evolution of resistance will not lead to the elimination of the phage from the habitat* (Campbell, 1961; Levin et al., 1977).

Unilateral selection in the bacterial population can actually augment the density of the phage population and possibly stabilize its association with the phage. This can be seen most readily by inspection of the equations for the equilibrium density of bacteria and phage. As long as the bacteria and phage can maintain their populations in the habitat, there will be an equilibrium with

$$\hat{P} = \frac{\psi - \varrho}{\partial}$$

$$\hat{N} = \frac{\varrho}{\partial(\beta e^{-\varrho\theta} - 1)}$$

(Levin et al., 1977). Thus, partially resistant bacteria are likely to have a selective advantage in cultures with phage and sensitive bacteria, and their evolution would result in an increase in the equilibrium density of the phage. The net effect of this would be a community that is further from the inelastic boundaries of $N = 0$, $P = 0$.

Selection in the phage population is necessarily antagonistic to the bacteria and can lead to the demise of the phage population. This, too, can be seen by an examination of the above equilibrium equations. Selection in the phage population would favor increases in the adsorption parameter δ and burst size β and reductions in the latent period θ. The effect of these changes in the infection parameters is a reduction in the equilibrium density of the bacteria. An increase in the adsorption rate parameter δ would also reduce the equilibrium density of the phage. The interested reader may wish to contrast the expectations derived from this model with those based on the predator–prey model presented in Chapter 3 by Roughgarden.

COEVOLUTION IN VIRULENT PHAGE AND THEIR HOSTS: EMPIRICAL CONSIDERATIONS AND EXTENSIONS

Resistance and persistence

From a priori considerations, we anticipate that selection in the bacterial and phage populations is antagonistic. The bacteria would be selected for resistance (reductions in δ), whereas the phage would be selected for higher levels of virulence (increases in δ and β and reductions in θ). It is clear that the potential for this type of antagonistic coevolution exists. One can readily isolate bacterial mutants that are fully resistant to phage to which other members of their clone are sensitive, and it is frequently possible to isolate *host range* phage that can attack these resistant bacteria as well as the sensitive cells. Indeed, bacterial resistance to phage was the phenotype used in the original demonstration of the randomness of mutation (Luria and Delbrück, 1943), and host range phage mutants played a very significant role in the early studies of bacteriophage genetics (Luria, 1945; Hershey, 1946).

This antagonistic selection in bacteria and their virulent phage has been observed in the various studies that have been done with experimental populations (Paynter and Bungay, 1969; Horne, 1970; Levin et al., 1977; Chao et al., 1977). These studies used a variety of different strains of *E. coli* and species of T phage, but in most cases phage-resistant mutants evolved and became the dominant clones. The evolution of these resistant mutants changes the continuous culture populations from a phage-limited to a resource-limited state. In the study by Chao et al. (1977) with *E. coli* B and the phage T7, there were at least three bacterial clones: the original sensitive, one mutant resistant to the original phage, and another mutant resistant to the original phage and to a host range mutant of that phage. At least two phage clones were present in these cultures: the original clone and a host range mutant of that clone. In spite of the antagonistic coevolutionary changes in these experimental populations, they persisted for extended periods of time (more than 80 weeks in the Horne, 1970, study).

It is of particular interest to ask why these "predator–prey" systems are stable (*sensu* persistence). If the bacteria are selected for a resistant mutant and if that mutation is not countered by a host range phage, why would the phage population not be eliminated? In the absence of mutants that are resistant to the phage and with continuous selection for higher levels of virulence, why would the phage not eliminate all of the bacteria? There are a variety of mechanisms that can account for the observed stability of the phage/host system. As demonstrated in a theoretical study by Levin et al. (1977), there are parameter values that specify stable states of co-existence for sensitive bacteria and virulent phage in the absence of genetic changes in their populations. Once one allows for evolutionary changes in these populations, there are at least three ways the association can continue to persist: (1) continuous selection for resistant hosts and counterselection for host range phage; (2) lower competitive performance of the resistant bacteria (relative to sensitive ones), and lower competitive performance of host range phage (relative to wild type) on sensitive hosts; and (3) the evolution of partially resistant bacteria with rates of adsorption that are still high enough to support the population of virulent phage. We believe that the second and third mechanisms are the primary ones accounting for both the short-term and long-term persistence of the associations between virulent phage and bacteria.

As demonstrated in theoretical studies by Campbell (1961) and Levin et al. (1977), sensitive bacteria and virulent phage can maintain a stable association as long as the phage-resistant cells are at a competitive disadvantage to sensitive cells in the absence of phage. This was observed in the experimental portion of the study by Levin et al. (1977). The introduction of T2-resistant clones of *E. coli* K-12 into chemostat populations of T2-limited *E. coli* B resulted in the ascent of the K-12 clone, with the persistence of the phage and of the sensitive *E. coli* B population. In phage-free competition, the T2-resistant K-12 clone was at a selective disadvantage relative to the T2-sensitive *E. coli* B. A similar result was obtained by Chao et al. (1977), but in that case T7-resistant clones of *E. coli* B evolved in that culture. Both the first-order resistant and second-order resistant *E. coli* in these experiments were, in the absence of phage, at a competitive disadvantage to the sensitive cell population from which they were derived. Chao et al. also found that the host range phage had a selective disadvantage when competing with the wild-type T7 for sensitive hosts.

Constraints on antagonistic selection

From these theoretical and empirical considerations, we postulate that to a great extent the evolutionary stability of phage/host associations can be accounted for by costs associated with resistance and host range shifts. Based on physiological considerations, one would expect that phage resistance imposes a cost on the bacteria (see discussions of constraints in Chapter 2 by Slatkin and in Chapter 3 by Roughgarden). The receptor sites to which the phage adsorb are membrane organelles that are likely to have other functions. Changes in their structure associated with resistance could impair these functions. In addition to the study by Chao et al. (1977), other evidence supports the view that mutations to phage resistance impose a cost. Demerec and Fano (1945) performed 50 pairwise competition experiments with a phage-sensitive clone of *E. coli* B and various mutants that were resistant to one or more T phage. In 34 of these phage-free experiments, the ratio of sensitives to resistants at the end of the experiment exceeded that at the beginning ($p < 0.05$). An analogous physiological argument could be put forth for the anticipated competitive disadvantage of host range phage relative to wild type, but perhaps more appealing is an a posteriori argument. If host range phage were as fit as or fitter than wild-type phage when competing for sensitive clones, there would be no wild-type phage. Unfortunately, save for the limited study of Chao et al. (1977), we are unaware of any experimental analyses of the relative competitive abilities of wild-type and host range phage. If there is generality to the observations of lower competitive ability for resistant bacteria and for host range phage and if physiological constraints prevent these disadvantages from being readily overcome, then a continuous progression of resistant and host range mutations is not only unnecessary for stability, but unlikely.

In addition to mutations that render bacteria absolutely resistant to phage, there are also those that result in quantitative reductions in the rate of phage adsorption. Although little consideration has been given to these "partially resistant" mutants, both theoretical considerations and informal results suggest they may play an important role in the evolution of stable host/virulent phage associations. One class of partially resistant mutants of *E. coli* excretes a mucilaginous substance (giving their colonies an aesthetically unappealing character). When cultures of *E. coli* K-12 (but not *E. coli* B) are challenged by a variety of different virulent and temperate phage, these mucoid colony types appear in high frequency among the surviving cells.

Cultures of these mucoid cells are able to maintain high density populations in the presence of phage and at the same time support a high density of the phage (B. R. Levin and P. Gidez, unpublished observations). We attribute this to a reduction in the rate parameter of phage adsorption associated with the mucoid phenotype. If these partially resistant mutations have only a small effect on cell growth rate, then their presence could preclude the evolution of fully resistant types. In that way, they might also restrict the evolution of host range mutants.

COEVOLUTION IN TEMPERATE PHAGE AND THEIR HOSTS: A PRIORI CONSIDERATIONS

A model

A schematic representation of the association between populations of temperate phage and host bacteria is shown in Figure 2. Uninfected cells are multiplying at a rate ψ_N, exclusive of losses to the phage and washout. As with virulent phage, temperate phage adsorb to host cells at a rate that is the product of cell density N and to the adsorption parameter δ. In contrast to virulent phage, however, not all infected cells are subject to cessation of growth and lysis. Instead, some frac-

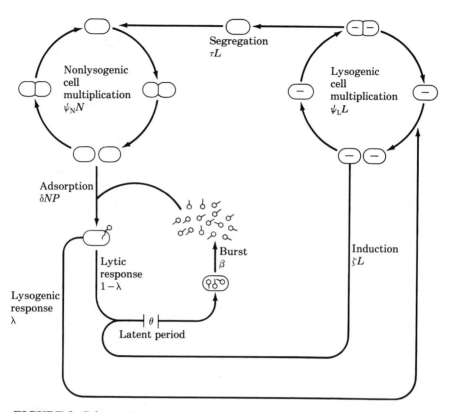

FIGURE 2. Schematic representation of the dynamics of a temperate phage and its bacterial host. N, L, Densities of nonlysogenic and lysogenic cells, respectively; ψ_N, ψ_L, rates of cell multiplication; λ, probability of lysogenic response given adsorption; ς, rate of induction; τ, rate of segregation; P, δ, θ, β, ϱ, see Figure 1. This model can be expressed as time-delay differential equations:

$$dN/dt = \psi_N N + \tau L - \delta NP - \varrho N$$
$$dL/dt = \psi_L L + \delta \lambda NP - \tau L - \varsigma L - \varrho L$$
$$dP/dt = \beta e^{-\varrho\theta}[\delta N'P'(1 - \lambda) + \varsigma L'] - \delta NP - \varrho P$$

where N', L', and P' refer to population densities θ time units before the present. (After Lwoff, 1953.)

tion λ of the infected cells incorporate the phage genome as a *prophage*. These *lysogenic* cells of density L continue to multiply at a rate ψ_L, which may differ from the growth rate of the nonlysogenic cells. The remainder $(1 - \lambda)$ of the infected cells exhibit the lytic response, just as if the adsorbed phage were virulent.

The lysogenic cells are immune to subsequent infection by phage particles of the same type. (At high levels of superinfection, this immunity may break down. This effect is not present in our model but has been considered in a model by Noack, 1968.) However, at a rate ς,

108

lysogenic cells are induced to exhibit the lytic response and thus enter the fated cell population. In addition, some lysogenic cells lose the phage genome through segregation at a rate τ, thereby entering the nonlysogenic cell population. As in the previous model, cells and phage are washed out of the habitat at a constant rate ϱ.

Let us again consider the consequences of changes in the various parameters of this model on the growth rates of cell and phage populations. In this case, however, we cannot view selection as acting independently on the sensitive cell, lysogenic cell, and free phage populations. Both the bacterial and phage genomes exist in two states, the former as sensitive cells and lysogens and the latter as prophage and free phage. Thus, in examining selection in this system, it is necessary to consider the phage and bacterial genomes at large rather than separately treat the different states in which they exist.

Selection on the bacterial genome

As in the previous model, any increase in the rate of cell multiplication is favored, for both lysogenic and nonlysogenic cells. Selection acting on the bacterial genome is expected to increase the likelihood of lysogeny (given adsorption), because any infected cell that does not become a lysogen is fated to death. Similarly, selection acting on the bacterial genome should minimize the rate of induction ζ, because induced cells are also fated to lysis.

Selection acting on the bacterial genome is somewhat more complex with respect to the parameters δ and τ. The direction of selection on the segregation rate will depend on the relative growth rates of lysogenic and nonlysogenic cells and on the relative death rates due to lysis of the two cell populations. If $\psi_L - \zeta > \psi_N - \delta P(1 - \lambda)$, then the phage genome is an advantage to its host and selection acting on the bacterial genome should minimize τ. Conversely, if the expected net growth of the nonlysogenic population is greater, then τ should be increased by selection acting on the bacterial genome. As with virulent phage, selection on the host genome should tend to reduce the rate of phage adsorption δ, *unless* the net rate of lysogenic cell multiplication is sufficiently greater than that of nonlysogens to offset the risks of lysis.

Selection on the phage genome

As with virulent phage, selection in temperate phage will be intense for increased rates of adsorption δ, because phage cannot reproduce outside their hosts. Similarly, the burst size β will be maximized and the latent period θ minimized.

109

The nature of selection acting on the phage genome for the likelihood of the lysogenic response λ is clearly critical to our understanding of the adaptive value of a temperate mode of existence (a $\lambda = 0$ renders a phage virulent). Therefore, it is important to examine the relative contributions of adsorbed phage that exhibit the lysogenic and the lytic responses. After time θ, the lytic response yields β free phage, whereas the lysogenic response nets $e^{\psi_L \theta}$ prophage (cells growing at rate ψ_L for time θ, assuming ζ and τ are near zero). Given realistic values for β (e.g., 100), θ (e.g., 0.5 hours), and ψ_L (e.g., 0.7/hour), the *short-term* dynamics are such that the lytic contribution will be far greater than the lysogenic contribution. However, if we compare the fates of the *progeny* free phage and the *progeny* prophage, we obtain a different conclusion. If nonlysogenic host cells are very rare, then free phage produced via the lytic response will have very low rates of subsequent reproduction, because adsorption to a new host is infrequent. In contrast, the progeny prophage resulting from lysogenic cell multiplication do not need to find a new host and can continue to multiply at the modest cellular rate indefinitely. If the product δN is sufficiently small to offset the short-term advantage of the lytic response, then selection acting on the phage genome should increase the probability of lysogeny (see also Campbell, 1961). *Thus, from the perspective of the phage, temperance appears to be an adaptation to low host-cell densities.*

Selection acting on the phage genome with respect to the rate of lysogen induction ζ will be of opposite direction and similar intensity to selection on λ, because induction is essentially a reversal of lysogeny. Under all conditions, selection in the phage population would be to minimize the rate of prophage loss by vegetative segregation, that is, to minimize τ.

If conditions are such that selection has favored phage temperance, an opportunity exists for the development of a mutualistic relationship with its host. As long as the cost in lower β or δ or higher θ is small, it would be to the advantage of the phage to carry genes that enhance the growth rate of lysogenic cells.

COEVOLUTION IN TEMPERATE PHAGE AND THEIR HOSTS: EMPIRICAL CONSIDERATIONS AND EXTENSIONS

Resistance and immunity

As with virulent phage, selection in temperate phage and their hosts *could* be antagonistic, that is, for more resistant hosts and more virulent phage. In accord with the model, selection would favor hosts that are resistant to the phage; but as long as the rate of mutation to resistant types is less than the probability of lysogeny, immune lysogens are likely to precede resistant clones. When one challenges sensitive *E. coli* with the temperate phage Lambda, the vast majority of surviving cells are Lambda lysogens rather than Lambda-resistant mutants. Indeed, Lambda-resistant clones are most readily isolated by challenging the sensitive bacteria with virulent mutants of Lambda. Thus, although resistance can evolve, the primary evolutionary response to infection with temperate phage should be the rise of the lysogenic population in which selection on the phage and bacterial genomes could be complementary.

At this time, we are aware of only one experimental study that has been directed at the evolutionary response of populations of sensitive bacteria to infection by temperate phage (J. Arraj and B. R. Levin, unpublished). The results of that study, with *E. coli* K-12 and Lambda in chemostats, support the hypothesis that the primary response of the bacteria is the rise of a lysogenic population. However, these results also suggest that the situation is somewhat more complex. Following the rise of Lambda lysogens, clones that are both lysogenic *and resistant* to Lambda appear and achieve substantial frequencies. It may seem redundant to be resistant as well as immune; however, immunity becomes ineffective at high levels of superinfection, whereas resistance apparently does not. The evolutionary importance of resistant lysogens has not yet been explored.

Selection for prophage-determined host phenotypes

In accord with the view that the direction of evolution in temperate phage is toward a mutualistic association with their host, there are a variety of prophage-determined characters that augment the growth rate of their hosts. The most obvious of these is, of course, immunity to subsequent infection by phage of that type, that is, *superinfection* immunity. There are also restriction enzymes coded for by prophage (Arber and Linn, 1969) and prophage-borne resistance to antibiotics (Williams Smith, 1972). Although the mechanism is not immediately apparent, we would anticipate that the diphtheria toxin coded for by the Beta phage of *Corynebacterium diphtheriae* (Uchida et al., 1971) enhances the fitness of the bacterium.

There is some evidence suggesting that prophage genes that augment host fitness are quite general. In a series of pairwise competition studies with *E. coli* in chemostats, Edlin and his colleagues have shown that under some culture conditions lysogens have a competitive advantage over nonlysogenic, resistant cells. They have obtained this result for a variety of different phage: Lambda, P1, P2, and Mu (Edlin et al., 1977; Lin et al., 1977). At this time, it is not clear how these prophage enhance the competitive performance of their hosts.

To us, the most intriguing problem concerning coevolution of bacterial viruses and their hosts is the conditions under which selection favors a temperate rather than a virulent mode of replication. We see three distinct hypotheses: (1) lysogeny enhances the stability of the phage–host association; (2) lysogeny enhances the fitness of the bacteria; and (3) lysogeny is an adaptation to low host densities. Although these hypotheses are not mutually exclusive, we consider the last mechanism to be the most important. We expand on this below.

It seems likely that the associations between temperate phage and their hosts are more stable than those of virulent phage and their hosts, that is, temperate phage are unlikely to drive their hosts to extinction. A number of authors have suggested differential extinction as a mechanism to account for the evolution and maintenance of temperance (e.g., Dove, 1971; Echols, 1972). This mechanism requires that selection operate at the level of the group (interdemic selection), the more stable phage–host associations having a group advantage. However, if virulent phage have an advantage *within* these groups, then the conditions for interdemic selection to favor temperate phage are likely to be very restrictive (Levin and Kilmer, 1974). If temperate phage have an advantage over virulent phage within these groups, then interdemic selection is not necessary.

If the prophage code for characters that enhance the fitness of their hosts, then selection could favor higher probabilities of lysogeny and lower rates of induction, that is, evolution in the direction of greater temperance. However, it seems unlikely that such "niceness" could be the *primary* selective pressure leading to lysogeny. For phage to express genes that enhance the fitness of their hosts, the phage genome would have to be maintained by host cells, that is, form some sort of prophage. Thus, from the perspective of the phage, they would already have to *be* somewhat temperate to *become* temperate. From the perspective of the host, selection would, of course, favor any mechanism that reduces the likelihood of a lytic infection by phage, even if that mechanism results in the maintenance of the phage genome. However, it is unlikely that the evolution of the temperate phage was through unilateral selection in the host population; although there is variation in the probability of lysogeny and rate of induction among bacteria, the proteins involved in insertion and excision (lysogeny and induction) are coded for by phage genes and *not* host genes.

The hypothesis that the temperate mode of phage existence evolved as an adaptation to low densities of sensitive cells seems, to us, the most parsimonious of the three. When sensitive hosts are rare, free phage produced via lytic infections would have low rates of subsequent reproduction, because adsorption to a new host is infrequent. On the other hand, prophage replication does not require a quest for new hosts. Although we know of no evidence to either support or refute this hypothesis, it should be amenable to direct experimental tests.

COEVOLUTION IN PLASMIDS AND THEIR HOSTS: A PRIORI CONSIDERATIONS

A model

In Figure 3 we present a schematic representation of the association between populations of conjugative (i.e., self-transmissible) plasmids and their host bacteria. The model presented in the figure is identical to that employed by Stewart and Levin (1977). The plasmid-free (P-F) and plasmid-bearing (P-B) cells grow at rates ψ and ψ_+, respectively.

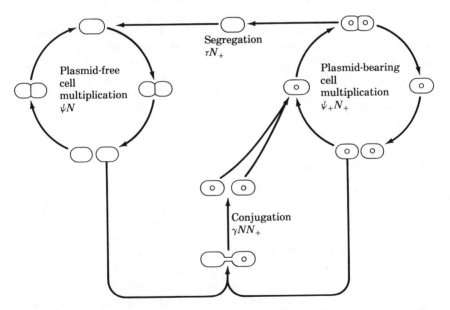

FIGURE 3. Schematic representation of the dynamics of a conjugative plasmid and its bacterial host. N, N_+, Densities of plasmid-free and plasmid-bearing cells, respectively; ψ, ψ_+, rates of cell multiplication; γ, conjugational transfer rate parameter; ϱ, τ, see Figures 1 and 2. This model can be expressed as differential equations:

$$dN/dt = \psi N + \tau N_+ - \gamma NN_+ - \varrho N$$
$$dN_+/dt = \psi_+N_+ + \gamma NN_+ - \tau N_+ - \varrho N_+$$

113

P-B and P-F cells encounter one another at random with a frequency that is proportional to the product of their densities. The proportion of encounters resulting in the transfer of a copy of the plasmid from a P-B cell to a P-F cell is governed by the conjugational transfer rate parameter γ. P-B cells can lose the plasmid, and thereby enter the P-F population, via vegetative segregation, at a rate τ.

Selection on the host genome

For both P-B and P-F cells, selection should favor an increase in the exponential growth rates ψ and ψ_+. The intensity of selection on the two growth rates will be independent of densities and relative frequencies of the component populations. Selection acting on the host genome will favor an increase in the segregation rate parameter τ if and only if the growth rate of the P-F cells exceeds that of the P-B cells.

With respect to the conjugative rate parameter γ, selection acting on the host genome is somewhat more complex. Any mutation arising in the P-B cell genome that increases *donor ability* would probably be selected against, because there is likely to be some cost associated with the mechanics of conjugative transfer (e.g., plasmid replication and the synthesis of structures called conjugative pili that are required for conjugation). This prediction is independent of the relative magnitudes of the growth rates of the P-B and P-F cells. However, a mutation that increases the *receptivity* of P-F cells will be favored if the plasmid augments the growth rate of P-B cells. The intensity of this selection for the recipient's contribution to γ will depend on the density of P-B cells.

Selection on the conjugative plasmid genome

Plasmid-borne genes that increase the growth rate or decrease the death rate of cells carrying that plasmid (those determining a higher ψ_+) would be favored by selection. In this model, the intensity of the selection for plasmid-borne genes that enhance host fitness would be independent of the frequency of P-B cells and the density of the population. Under all conditions, selection would favor plasmid-borne genes that augment the stability of the plasmid in the host, that is, that reduce the segregation rate τ. The intensity of this selection would be independent of the density of the population or the frequency of P-B cells.

Selection would favor plasmid-borne genes that increase the rate of infectious transfer of the plasmid. The intensity of this selection for higher γ would be directly proportional to the population density and the frequency of P-F cells. At very low densities of P-F cells, infectious transfer would make a negligible contribution to the rate of increase of the plasmid. Thus, if there were a significant cost associated with conjugative ability, selection in the plasmid could actually favor reductions in the rate of infectious transmission when recipients are rare.

Selection on the nonconjugative plasmid genome

From one perspective, nonconjugative (non-self-transmissible) plasmids can be considered as a limiting case of the conjugative factors (where the rate constant of transfer $\gamma = 0$). However, this simple interpretation is not really sufficient. Nonconjugative plasmids, like segments of the host chromosome, can be infectiously transmitted by being "picked up" by either conjugative plasmids or phage, a process known as *mobilization*. In some cases mobilization can be quite effective (e.g., Levin and Rice, 1980).

In modeling the population biology of nonconjugative plasmids, one must consider the population dynamics of the mobilizing replicon(s), as well as that of the nonconjugative plasmid. This is an exercise we shall refrain from in this forum (but see Levin and Stewart, 1980). The primary issue of concern here is that even for non-self-transmissible plasmids, there could be selection for changes in the rates of infectious transfer by mobilization. The nature and direction of this selection would be similar to that considered for the rate constant of conjugative plasmid transfer.

Plasmids coding for allelopathic substances

In the model presented in Figure 3, the plasmids may increase the fitness of cells by augmenting the growth rate of their immediate hosts. There is, however, an important class of plasmid-determined characters for which the plasmid is to the *disadvantage* of the individual host carrying it but which enhance the fitness of the P-B population at large. The most extensively studied plasmid-determined characters of this type are the *bacteriocins* (Reeves, 1972). These are proteins that kill sensitive bacteria of the same or closely related species. In addition to coding for the production of these *allelopathic* molecules, bacteriocinogenic plasmids also confer immunity to these agents. The individual cells carrying these plasmids are at a disadvantage because bacteriocin synthesis and release is lethal to the host cell. At any given time, however, only a small minority of the bacteriocinogenic population is induced to synthesize and release bacteriocin. If there is a cost associated with the carriage of plasmids coding for other antibiotics (such as those produced by the streptomycetes; Hopwood and Merrick, 1977), then these too fall into this class of plasmid-determined phenotypes.

115

Although the model presented in Figure 3 is not an accurate analog of the population biology of allelopathic plasmids, the major components of the population dynamics of these types of replicons have been mimicked by simple models (Chao, 1979). The primary conclusion drawn from Chao's model of nonconjugative bacteriocinogenic plasmids is that selection for these elements would be frequency-dependent. Cells carrying them could only increase when they are relatively common. The reasons for this are rather straightforward. Due to the lethal synthesis and other costs associated with these factors, bacteriocinogenic cells would be at a disadvantage when competing with sensitive P-F cells, *unless* the concentration of that allelopathic agent is high enough to kill sufficient numbers of P-F (i.e., sensitive) cells to make up for that competitive disadvantage. The latter will occur only when the P-B cells are at relatively high densities.

COEVOLUTION IN PLASMIDS AND THEIR HOSTS: EMPIRICAL CONSIDERATIONS AND EXTENSIONS

Selection and persistence: plasmids are not just "selfish DNA"

Plasmids, unlike phage, have no free state; they are parts of cells, like chromosomes. Thus, one might conclude that their evolution would be toward an increasingly mutualistic relationship with their hosts. There are, however, possible exceptions to this mutualistic form of coevolution. If the plasmids impose a cost on their hosts (i.e., reduce their rates of growth), selection in the plasmid and host populations would be antagonistic. The plasmids would be selected for higher rates of infectious transmission and the hosts for resistance to infection by plasmids and higher segregation rates. Thus, in considering the nature of coevolution in plasmids and their hosts, it is first necessary to ask (1) whether plasmids are likely to impose a fitness cost on their hosts and (2) whether plasmids can be maintained by infectious transfer alone.

In the absence of selection favoring plasmid-borne genes, cells carrying plasmids have a competitive disadvantage relative to identical cells without those extrachromosomal genetic elements. This is, in fact, what one would anticipate on physiological grounds. The additional DNA and protein synthesis associated with the carriage of plasmids must impose some cost on a bacterium. The magnitude of this cost for conjugative plasmids is surprisingly high. In competition experiments between P-B and P-F *E. coli* in conditions where there was no selection for plasmid-borne genes, Levin (1980) reported a 10% lower growth rate for cells carrying the conjugative plasmid R1. Whether plasmids generally impose this high a cost on their hosts remains to be seen.

Using a mathematical model like the one presented here for conjugative plasmids, Stewart and Levin (1977) demonstrated the existence of a broad set of conditions under which infectious transfer can overcome segregational loss and substantial levels of selection against P-B cells and thereby lead to the maintenance of the plasmid in high frequency. Although these conditions for the maintenance of plasmids deleterious to their hosts can be met under laboratory conditions (with high density cultures and mutant plasmids with very high transfer rates), Levin et al. (1979) and Levin (1980) suggest that these conditions are unlikely to obtain in natural populations of bacteria. This idea can be stated as a hypothesis: *for plasmids to be maintained in natural populations of enteric bacteria, they must carry genes that (under at least some conditions) enhance the fitness of their immediate hosts or that of cells carrying that plasmid in the population at large.* "There are no neutrals there" (Reece, 1932), that is, plasmids are not just "selfish DNA" (Doolittle and Sapienza, 1980).

If the above hypothesis is valid and general, then for all plasmids and their hosts, coevolution would necessarily be mutualistic. The existence of a diverse array of plasmid-determined phenotypes is clear evidence in support of this interpretation. Most characters coded for by plasmid-borne genes can enhance the fitness of the bacteria carrying that element. Just how significant plasmid-determined characters are for bacterial adaptation is very dramatically (and from a clinical perspective, frighteningly) demonstrated by the rise of antibiotic resistance. Most clinically important antibiotic resistance in bacteria is determined by plasmid-borne genes; and since the start of the antibiotic era in the late 1940s, the frequency of bacteria carrying antibiotic resistance (R) plasmids has increased enormously. At present, the majority of clinical isolates for some pathogenic bacteria like *Salmonella typhimurium* (gastric enteritis) and various species of *Shigella* (bacterial dysentery) carry at least one R-plasmid (see reviews by Anderson, 1968; Mitsuhashi, 1971; Falkow, 1975).

In addition to the increase in the frequency of bacteria carrying R-plasmids, the number of antibiotic resistance genes carried by single R-plasmids also increased during this period. When these plasmids were first discovered in the late 1950s (Watanabe, 1963), the majority isolated from enteric bacteria conferred resistance to one or two antibiotics. The average number of resistances determined by single R-plasmids increased rapidly during the early 1960s (Anderson, 1968). Currently, it is not uncommon to isolate from enteric bacteria R-plasmids that have four or five resistance genes.

The rapid rate by which plasmids acquire additional antibiotic resistance genes can be attributed to the fact that these genes are often

117

present as parts of transposable genetic elements (i.e., transposons; Falkow, 1975; Broda, 1979; Campbell, 1981). Genes on transposons do not require sequence homology for recombination and insertion into chromosomes and plasmids. In the course of their travels among different hosts, conjugative plasmids pick up transposons with different resistance genes. As long as bacteria are confronted with a variety of different antibiotics, plasmids that confer resistance to more antibiotics would have a selective advantage.

In accord with this hypothesis of mutualistic coevolution, we would anticipate that if plasmids augment the fitness of their hosts, there would be selection on the host to increase the receptivity to plasmids and reduce the rate of segregational loss. At this juncture, we know of no evidence to either support or refute this interpretation. However, because one can select hosts that are refractory to conjugative plasmids (Reiner, 1974), it should be possible to test this hypothesis.

The allelopathic plasmids

In our a priori considerations, we suggested that selection acting on plasmids that code for allelopathic substances would be frequency dependent. The results of studies that have been done with colicinogenic plasmids in experimental populations support this interpretation (colicins are bacteriocins that affect *E. coli* and closely related species). Using a variety of different Col plasmids, Zamenhof and Zamenhof (1971), Adams et al. (1979), Chao (1979), and Chao and Levin (1981) did competition experiments with colicinogenic and sensitive *E. coli* in chemostats. In all of these experiments, selection favored the colicinogenic cells only when they had initial frequencies in excess of 1%.

Because the sensitive bacteria are being killed by the colicin, it seems reasonable to assume that in these experiments, there would be selection for colicin-resistant cells. This was, in fact, observed in the colicin E1 study of Adams et al. (1979). Chao and Levin (1981) also reported the existence of mutants that were resistant to the colicin E3 they were studying, but in their cultures, these resistant cells did not achieve substantial frequencies. They attributed this to a marked reduction in competitive performance associated with the Col E3 resistance mutation.

The importance of physically structured habitats

Frequency-dependent selection for allelopathic plasmids raises a question about the evolution and maintenance of these types of plasmids. If they cannot increase in frequency *until* they are relatively common, how does one account for their evolution and persistence? One possible explanation is "hitchhiking," that is, bacteriocin-determining genes

118

are on plasmids that code for characters that enhance the competitive performance of the individual cells carrying them (e.g., antibiotic resistance). An alternate hypothesis, which we favor, was suggested by Reeves (1972) and expanded upon and tested by Chao (1979) and Chao and Levin (1981). According to this hypothesis, bacteriocins (and, by extension, other allelopathic substances like antibiotics) are adaptations for interference competition (Gill, 1974) in *physically structured* habitats. Although cells carrying the Col E3 plasmid could not increase when rare and when competing with P-F sensitive cells in liquid (mass) culture, they could increase when competing in a soft agar matrix.

In a physically structured habitat, the bacteria grow as colonies rather than as individual cells (as they do in mass culture). The allelopathic agents synthesized by induced cells of bacteriocinogenic colonies diffuse out into the environment and kill sensitive cells in the vicinity of that colony. The net effect is reduced competition for limiting resources for colonies of cells carrying plasmids coding for allelopathic substances, and the production of more P-B cells. As a result of this *resource sequestering*, cells carrying allelopathic plasmids have an advantage over sensitive cells at all *frequencies* (although only at high *densities*, where competition is important). Presumably, this same type of mechanism would also favor colonies that are lysogenic for temperate phage.

Constraints on infectious transmission

Because it is copies of plasmids rather than the plasmids themselves that are transmitted by conjugation, it is reasonable to assume that selection would always favor plasmids with higher rates of infectious transmission. But empirical considerations suggest that this is not the case. There are a vast number of apparently viable plasmids that are non-self-transmissible, and among these there are some that are extraordinarily difficult to mobilize with conjugative plasmids. Moreover, most conjugative plasmids isolated from natural populations are repressed for conjugative pili synthesis (Meynell, 1973). At any given time, only a minority of the P-B cell population produces conjugative pili and is capable of transmitting the plasmid. Shortly after the receipt of a plasmid, a cell produces conjugative pili and is capable of transmitting that element. As time proceeds, a plasmid-coded repressor protein accumulates in that cell and its descendants and competence for transfer of that plasmid declines. For the wild-type (repressed) plasmid R1 in steady-state P-B populations, the rate constant of plasmid transfer γ is three orders of magnitude lower than that for its permanently derepressed mutant, R1-drd-19 (see Levin et al., 1979).

We see two (not mutually exclusive) hypotheses for the evolution of repressible conjugative pili synthesis and the generally lower-than-possible fertility of wild-type plasmids. Anderson (1968) suggested that repression of conjugative pili synthesis is a mechanism to avoid infection by donor-specific bacteriophage that adsorb to conjugative pili. The second hypothesis, which we prefer on grounds of parsimony and generality, asserts that repressible conjugative pili synthesis and the existence of non-self-transmissible plasmids is a consequence of the limited opportunities for transfer in natural populations and the high cost of self-transmissibility. As we pointed out in our a priori considerations, the intensity of selection for infectious transmission is dependent on the density of potential recipients. If the latter is low, as a result of a low overall population density or a low relative frequency of possible recipients, and if there is a cost associated with infectious transmission, selection would favor lower rates of infectious transmission.

The mechanisms for conjugative pili synthesis require a substantial number of genes: approximately 15 megadaltons of DNA or approximately ¼ of the genome of a plasmid like R1 (Willetts, 1972). As a result of this additional DNA and the synthesis of conjugative pili, the burden imposed by conjugative plasmids would even be greater than that for nonconjugative elements. The results of the limited number of experiments we have done on this problem are consistent with this view. In competition experiments with the nonconjugative plasmid pCR1, P-B cells had a disadvantage of less than 5% (relative to P-F cells) as compared to 10% for the conjugative plasmid R1 (Levin et al., 1979). Moreover, cells carrying the permanently derepressed mutant, R1-drd-19, had a disadvantage of between 15 and 20% relative to P-F cells of their type (Levin, 1980). Thus, unless the density of recipients is substantial, one would not anticipate selection to favor higher rates of infectious transmission.

DEFENSE MECHANISMS AND MEASURES TO COUNTER THEM: A POSTERIORI EVIDENCE FOR COEVOLUTION

Compelling evidence for the long-term coevolution of plasmids, phage, and their hosts is the existence of specific systems to prevent and/or limit infections by parasitic DNAs and mechanisms to counter these barriers to their replication. Host defense mechanisms operate in three basic ways: (1) preventing novel DNAs from entering cells, or *exclusion*; (2) destroying novel DNAs that do enter, or *restriction*; and (3) preventing foreign DNA from replicating at a rate sufficient for persistence, or *incompatibility* (Bennett and Richmond, 1978). In the following section, we briefly consider the nature of these defense mecha-

120

nisms and how they serve as evidence of coevolution among these replicons and their hosts or, in some cases, among different replicons.

Exclusion

The cell envelope is probably the most important barrier to infections by plasmids and phage. To some extent this may be coincidental, perhaps like our "resistance" to Dutch elm disease. However, in other cases it is clear that an exclusion mechanism evolved for the specific purpose of limiting invasions by parasitic replicons. The existence of mutations conferring resistance to bacteriophage surely stands as evidence for the potential for the evolution of exclusion. Analogously, the existence of host range mutations serves as evidence for the potential of phage to evolve mechanisms to overcome cell envelope defense. Some exclusion mechanisms are determined by replicons themselves. Many conjugative plasmids code for mechanisms that prevent entry by closely related plasmids (Novick, 1969). It seems reasonable to assume that these evolved for plasmids to prevent competition for replication within individual bacteria.

Restriction

As is the case with the microparasites of multicellular organisms, once past their host envelope the parasitic DNAs of bacteria have to contend with "immune" systems of their hosts. Some of these intracellular mechanisms to prevent replication of invading DNAs are rather limited in the range of DNAs upon which they can operate, for example, the repressors responsible for superinfection immunity for temperate phage. In other cases, the range of foreign DNAs acted upon is quite broad. The *restriction modification* systems are perhaps the prime example of "immune" systems of the latter type (Arber and Linn, 1969; Stent and Calendar, 1978).

Novel DNAs entering a cell are cut at specific sequences of bases by *restriction endonucleases*. As is the case for the generalized "immune" systems of higher organisms, there is a need to distinguish "self" from "nonself." This is accomplished by modifying bases within the cleaving sequences recognized by the restriction endonuclease, by adding methyl groups to the cytosines or adenines in these cleaving regions. The modification methylases that catalyze the latter reactions are coded for by genes that are closely linked to those for the restriction endonucleases.

The effectiveness of restriction varies considerably among the vari-

121

ous phage and plasmid DNAs penetrating the cell envelope. In some cases, it is clear that restriction systems are very effective, reducing the likelihood of cells succumbing to a mortal infection by a phage by four or five orders of magnitude. However, as one might anticipate, the modification system necessary for recognizing self limits the effectiveness of restriction. Phage or plasmid DNA that evade restriction are modified and recognized as self DNA. These modified replicons are fully effective against bacteria with that restriction modification system. Some phage, such as T-even coliphage, have 5-hydroxymethyl-cytosine instead of cytosine as part of their normal genome and are therefore relatively immune to most restriction enzymes. Could it be that the atypical bases of these phage evolved as a mechanism to overcome host restriction?

Those investigators using specific endonucleases for DNA manipulations (e.g., gene splicing) have shown that there are many different restriction enzymes with different cleavage sites. These enzymes are coded for by host, plasmid, and temperate phage genes and are present in a very phylogenetically and ecologically diverse array of bacterial groups. Included among these is *Thermoplasma acidophilium* (McConnell et al., 1978), an Archaebacterium that lives in coal piles and is most readily cultured between pH 1 and 2 at 59°C (Searcy et al., 1981). If restriction modification systems are, as they appear to be, defense against novel DNAs, the *Thermoplasma* situation clearly indicates just how universal the problem of coping with parasitic DNA is to bacteria.

Incompatibility

The successful passage through the gauntlet of exclusion and restriction defenses does not ensure maintenance of a parasitic DNA in a bacterial lineage. For stable inheritance, it is necessary for that DNA to replicate at a rate at least as great as that of the host chromosome and, upon cell division, to be transmitted to both daughter cells. Temperate phage and plasmids have evolved a variety of mechanisms to ensure this vertical transmission. The incorporation into the host chromosome by prophage and the production of multiple copies by plasmids clearly augment the likelihood of vertical transmission. Many autonomous single-copy replicons are also very stable (with vegetative segregation rates of 10^{-5} per generation or less) and some rather extraordinary mechanisms have evolved to ensure their stability (see Austin et al., 1981).

Operating against the stable inheritance of replicons are a variety of incompatibility systems. In some cases, the invading DNA remains intact and expresses its genes but fails to replicate at all (Stocker, 1956; Hayes, 1968). In other cases, replication does occur, but the seg-

122

regation rate is high and the replicon, usually a plasmid, can only be maintained in a lineage by continuous selection for the genes carried by that element. In some cases these incompatibilities may be coincidental; there is no specific mechanism to preclude replication, but the replicon simply has not evolved a mechanism to allow for its stable inheritance. This may well be the case where the plasmid is transferred to a species that is phylogenetically very distant from the donor. There are, however, incompatibility systems that have clearly evolved as mechanisms to preclude the stable inheritance of invading replicons. The best studied of these are coded by plasmid rather than host genes and operate most effectively against plasmids of the same or closely related types (Falkow, 1975). Why do these and other replicon-determined defense mechanisms fit the cliché of competition being most intense among closely related species?

COEVOLUTION AND BACTERIAL SEXUALITY: MUCH ADO ABOUT VERY LITTLE

Save for the incorporation of free DNA (a process known as *transformation*), plasmids and phage are the sole vectors for the exchange of genetic material between bacteria. Thus, it might seem that a good deal of the coevolution of these organisms would be directed toward the role of plasmids and phage as vehicles of recombination. We suggest that this has not been the case. Although gene exchange mediated by these vectors is unquestionably important to bacterial adaptation and evolution (Reanney, 1976; Bennett and Richmond, 1978; Davey and Reanny, 1980), it is unlikely that natural selection has acted *directly* to increase the effectiveness of these replicons as vectors for recombination.

In natural populations of bacteria, recombination appears to be an extremely rare event from the perspective of an individual bacterium. To be sure, high rates of recombination can be obtained with laboratory strains such as *E. coli* K-12 and permanently derepressed F plasmids that incorporate into the host chromosome (*Hfr*: see Hayes, 1968). However, most naturally occurring plasmids are repressed for conjugative pili synthesis (Meynell, 1973) and do not readily incorporate into the chromosomes of their hosts (Holloway, 1979). Based on estimates of the rate constants of plasmid transfer and phage adsorption, on the likelihood that these vectors will pick up and transfer host genes, and on the densities of natural populations, Levin (1981) suggests that for *E. coli* in their natural habitat the per capita rate of gene exchange by plasmid- and phage-mediated recombination is as low as or lower than by mutation (10^{-6} per cell per generation or less). The results of electrophoretic studies of structural gene diversity in natural populations of *E. coli* (Selander and Levin, 1980; Caugant et al., 1981) are consistent with this interpretation. Gene complexes are maintained for extended periods of time without being broken down by recombination, and natural populations are far from linkage equilibrium.

The fact that in laboratory culture one can select for plasmid and phage that are effective vehicles for host gene recombination and that vectors of this type do not exist in natural populations is, of course, a posteriori evidence for the absence of selection for high rates of recombination in natural populations. This, however, begs the question of *why* high rates of phage- and plasmid-mediated recombination have not been selected for. We believe that part of this answer lies in the fact that being a vehicle for the transfer of host genes is likely to be a disadvantage for a plasmid or phage. This is clearly the case for some of the plasmids and phage used for recombination analyses in genetic studies. In the case of the F plasmid incorporated into the chromosomes of *Hfr*s, the F replicon is not transmitted in entirety until the whole host chromosome is transmitted (Hayes, 1968). Thus, that plasmid would usually gain neither from the advantages of being in the recombinant nor from the mobility of infectious transfer. For many general transducing phage, the individual virus responsible for recombination contains few or possibly none of its own genes. In the course of replication in the donor cell, it "accidentally" picks up a headful of host genes. Whether all high frequency recombination plasmids and phage used in laboratories are less fit than the native replicons from which they were derived remains to be seen.

A more general explanation for the low frequency of plasmid- and phage-mediated recombination arises from the limitations of natural selection for altering the frequency of occurrence of intrinsically rare events. This can be seen if we take the rather extreme view that the receipt of random genes necessarily augments the fitness of the recombinant (a view we would not want to defend), for example, increasing its growth rate by 10%. If the basal probability of recombination is 10^{-6} and if a mutation doubles the probability of a cell becoming a recombinant (or doubles the probability of a vector transmitting host genes), then for cells of the mutant type the expected growth rate is $\psi(1+2\times10^{-7})$ as compared to $\psi(1+1\times10^{-7})$ for the nonmutant type; this is a very low selective differential and one that is likely to be overridden by stochastic factors or periodic selection (Atwood et al., 1951; Levin, 1981).

In discussing the evolution of sex in bacteria, we have intentionally neglected group or interdemic selection. In fact, we do not believe *any* selection is necessary to account for plasmid- and phage-mediated recombination in bacteria. It is most parsimonious to assume that these low rates of recombination are the results of errors in replication and infectious transfer for the vector plasmids and phage. Although these errors are not products of natural selection, they play an important role in the adaptation and evolution of bacteria, perhaps approaching the significance of the errors responsible for mutation.

AN OVERVIEW

We have attempted to portray the various types of plasmids and phage as functionally similar genetic elements, that is, as parasitic replicons. Our intent was to deemphasize the differences in their modes of vertical and horizontal (infectious) transmission and offer a more unifying view of this phenomenon. However, because we succumbed to the convenience of separate treatment, we fear the reader may have failed to appreciate a more comprehensive interpretation. For this reason, some emphasis, explanation, and expansion seems warranted.

By suggesting that the various kinds of plasmids and phage are a single type of genetic element, we *do not mean* to imply that they have a common ancestry. This is clearly not the case for bacteriophage and unlikely to be so for plasmids. Bacteriophage have a variety of different genetic molecules: some single-stranded DNA, some double-stranded DNA, and some RNA; and these molecules replicate in a number of fundamentally distinct ways (Stent and Calendar, 1978). Although all known bacterial plasmids appear to be covalently closed circles of double-stranded DNA, this may be the result of convergent evolution due to the constraints of vertical transmission. There are at least two mechanisms of control for the replication of plasmid DNA (Falkow, 1975), and it may well be that the assay methods used may fail to detect plasmids of other types of genetic molecules. Because of these differences, we conclude that extant bacterial plasmids and phage are polyphyletic. Furthermore, it is likely that many plasmids and phage do not have unique ancestries but are chimeras composed of components from a variety of plasmid, phage, and host lineages.

It seems reasonable to suppose that all lineages of bacterial plasmids and phage were originally derived from the DNA or RNA of bacteria or possibly that of higher organisms and that incipient plasmids and phage are continually being generated from these sources. It also seems likely that some plasmids and phage evolved from each other. It is clear that the lines separating the different modes of replication and horizontal transmission are neither sharp nor insurmountable. Some

125

prophage, such as P1, replicate as plasmids (Ikeda and Tomizawa, 1968). With modest genetic changes, temperate phage can be made virulent (Lwoff, 1953; Ptashne, 1971), conjugative plasmids can be made nonconjugative (Willetts, 1972), and temperate phage can be made into plasmids (Signer, 1969).

Viewing the various types of plasmids and phage as a single type of genetic element is useful for evolutionary and ecological considerations. In their nascent phase, as they emerge from cellular DNA or RNA, all of these elements would be autonomous replicons without specific mechanisms to assure either their maintenance in the descendants of their original host cell or their infectious transmission. They are likely to be confronted with some physiological mechanism of selection by the host to rid itself of the burden of foreign DNA or RNA. To survive in this hostile climate, the incipient plasmid or phage would have to evolve some mechanism for "over-replication" (Campbell, 1981). We see two nonexclusive ways for this to occur: (1) "niceness," that is, acquiring genes that enhance host fitness; and (2) infectious transmission. As long as there is a finite rate of segregation, becoming innocuous (selectively neutral) and replicating with high fidelity would not be sufficient for maintenance.

The route that is taken for over-replication will depend on the genetic and physiological constraints on the replicon and its host and on the environment of the bacterial population. With respect to the latter, we believe that *population density of the host is the primary factor in determining the form of over-replication.* In high density populations of bacteria, replicons with effective mechanisms of horizontal transmission would be favored. The extreme of this would be virulent phage and the resulting antagonistic coevolution. As the density of the host population declined, so would the intensity of selection for infectious transmission. Niceness would become increasingly important for the persistence of the autonomous replicon, and mutualistic coevolution would result. In bacterial populations of intermediate density or in populations with high amplitude oscillations in density, temperate phage and conjugative plasmids would flourish. In populations with sustained low densities, the costs associated with the mechanisms for infectious transmission could not be overridden, and niceness would serve as the only means available for the over-replication of parasitic genetic molecules; under these conditions, nonconjugative plasmids would be favored.

We have made a number of general and specific statements about the nature and direction of coevolution in bacteria and their viruses and plasmids. Although somewhat legitimatized by mathematical modeling and selected facts, most of these statements about how things came to be are no more than microbial "just so stories." As is the case with other evolutionary phenomena, there is no way to formally demonstrate that the suggested pathways are indeed the actual ways things came to be. However, in the case of bacteria and their plasmids and phage, these evolutionary hypotheses can be readily tested with experimental and natural populations. We hope researchers will find some worthy of testing.

CHAPTER 6

ENDOSYMBIOSIS

Lee Ehrman

A fascinating variety of adaptations have developed by coevolution, from the nitrogen-fixing bacteria (*Rhizobium*) that live in specialized nodules on the roots of legumes to the nudibranch—a sea slug that starts life as a larva entrapped in jellyfish tentacles, is then engulfed by its jellyfish, but ends its cycle with the slug full grown and the medusa a tiny vestigial parasite residing on the ventral surface of the snail's mouth (Thomas, 1979).

These designs for living are produced by symbiosis—a condition in which organisms of different species live together in a state of mutual influence. There are several types of symbiotic relationships: parasitism, in which one species lives at the expense of the other; commensalism, where one species benefits from the association while the other is neither harmed nor benefited; and mutualism, where both species benefit from their relationship.

Broadly defined though they are, these affinities have many levels of integration. The continuum may range from nearly autonomous partners to components that are so merged that the identity and function of one of the organisms is difficult to ascertain. These patterns of association and the levels of integration of the partners provide valuable clues to the history of the coevolution of the symbionts and how their relationships have shaped their metabolic and behavioral interactions.

The most intimate of these coevolved systems is endosymbiosis, in which one of the organisms incorporates the other—an intricate cohabitation of two species in which a symbiont exists in the cells of a host for at least a discrete portion, if not all, of its partner's life cycle.

Margulis (1970) postulates that endosymbiosis explains the origin of the eukaryotic cell, which houses its genetic material in a nucleus. The serial endosymbiosis theory stipulates that the incorporation of free-living prokaryotes, represented now by the mitochondria and chloroplasts, by a host cell is the paradigm for the emergence of the eukaryotic cell and led to one of the most important phylogenetic distinctions—that between prokaryotes and eukaryotes. Thus, endosymbiosis is considered one of the great primeval events in our evolutionary history. Margulis (1976) notes that "the biosphere is conspicuous for the frequency and diversity of associations between organisms that share only remote ancestry. It can be argued in some cases that the origin of certain higher taxa was made possible by merging alliances able to perform functions that individual partners could not."

From an abundance of literature on endosymbiotic liaisons, I shall describe two cases that provide us with some insights into the evolutionary dependence of host and endosymbiont and the implications of these symbiotic systems as bravura evolutionary events.

BACTERIAL SYMBIONTS OF *PARAMECIUM AURELIA*

Among the ciliated Protozoa, the *Paramecium aurelia* complex (of 14 species) houses intracellular bacterial symbionts belonging to the genus *Caedobacter*. Some of these endosymbionts are characterized by the fact that they produce toxins. The paramecia that bear them are called killers. Sonneborn (1938) discovered the killer trait in *P. aurelia* and provided evidence that it was cytoplasmically inherited. Five years later Lindegren and Altenburg separately suggested the microorganismal nature of the killer trait. All were correct.

The inhabitants of *P. aurelia*, known as kappa, are now classified as *Caedobacter taeniospiralis*. These bacteria in turn often carry one or more viruses or bacteriophages. Thus, we have endosymbiosis with at least three participants. As the hosts and their endosymbionts have evolved, the relationships have become highly specific, so that the hosts are intolerant of the endosymbionts borne by other hosts. Quackenbush (1977) has proposed that *C. taeniospiralis* (kappa) comprises more than one species.

When two strains of paramecia of complementary mating types are mixed together and the result is death of one strain, the surviving strain can be assumed to harbor an agent (or symbiont) lethal to the moribund strain. The results of mixing two strains of paramecia vary according to the type of kappa, the type of phage, the sensitivity of the paramecia, and environmental conditions (Preer, 1975). Thus, even within a single species of paramecium, different genotypes carry different kinds of kappas, and a paramecium is sensitive to the strains of kappa carried by other genotypes.

In most species of the *P. aurelia* complex, the bacteria cannot infect

129

via the medium but are transmitted strictly by heredity (Preer, 1975). If a paramecium does not harbor kappa, it is termed "sensitive" and is vulnerable to kappa's toxin. Each genotype of *P. aurelia* is resistant to the kappa it may house, but other effects of kappa on its host are obscure.

Kappas employ glycolysis and respire (Kung, 1970, 1971). They contain citric acid cycle enzymes and possess cytochrome. Kappas have neither mitochondria nor nuclear membranes. The shapes and sizes of their cell walls resemble those of other bacteria, and like many other bacteria, they harbor lysogenic phages. Nearly half of the kappas are distinguished by the presence of refractile bodies (R bodies), which are proteinaceous, ribbon-like, tightly wound (but capable of unwinding) structures thought to be products of defective phages. It is believed that the phages housed by kappas specify the toxin (Preer et al., 1974).

In 3 of the 14 known species in the *Paramecium aurelia* complex, it has been shown that the paramecia–hosts excrete a poison of kappa particles into the culture media. For a paramecium to be toxic, it must inherit its cytoplasm from a killer strain with at least one nuclear dominant gene K, which provides tolerance to the kappa. The recessive allele k precludes the maintenance of kappa in the cytoplasm. Thus, tolerance can be bred out of resistant paramecia, because when Kk paramecia reproduce sexually, the issue may be KK (resistant) or kk (sensitive). However, if a KK paramecium does not actually maintain kappa in its cytoplasm, it will find the kappas in its food to be toxic. The kappas that had been liberated into the medium enter through the gullet, attack the food vacuoles, and induce blebs (blisters) on the cuticle. The presence of the ingested kappa with its toxin-producing phage proves lethal not only for kk paramecia, but for the kappaless K genotype as well. Thus, kappa cannot become established in a new host that does not already harbor kappa (J. Preer, personal communication).

Killer stocks of paramecia can be made sensitive through a number of laboratory techniques that remove their kappas: exposure to high temperatures, X rays, nitrogen mustard, chloromycetin, or inadequate media. Also, the paramecia can be induced to multiply faster than the kappas can; this procedure results in new paramecia that have not inherited any kappa. At least some of these "cures" must sporadically occur in nature, but the number of sensitive paramecia does not increase because of their resultant vulnerability.

As different species of bacterial symbionts with kappa-like activity have been discovered, other Greek letters have been used to identify them. *Mu* (mate-killers) is a species that cannot be transmitted via the medium but needs cell-to-cell contact to kill. *Lambda* and *sigma* are rapid lysis killers, usually showing their lethal effects on sensitive

130

paramecia in less than 30 minutes. *Delta* has been found in all paramecia known to carry endosymbionts. It was originally reported to be
a killer but does not show that property now in laboratory-maintained
stocks.

Preer (1977) considers these bacterial endosymbionts to be very ancient. New infections by free-living bacteria have never been observed
in the laboratory and transfer at conjugation has been shown to be
unlikely. Because symbiont-bearing cells contribute more progeny to
the next generations than cells that do not bear symbionts, it appears
that kappas confer advantages upon their host.

The prime one is via the inhibition of competitors for resources (see
Gill, 1974). Preer (1975) notes that "the presence of the symbiont
renders its host resistant to the toxin which it produces. The endosymbionts obviously profit from the association with their hosts, acquiring
from them all necessary nutrients, a place of abode and a buffer between them and the external environment . . . . The endosymbionts are
dependent on specific nuclear genes of *Paramecium* for their maintenance. These genes appear to be highly specific in respect to the particular endosymbiont whose maintenance they control."

The paramecium–kappa relationship is thought to be an ancient
one established before geological movements isolated today's continents (Preer, 1977). Their widespread distribution in North America,
Scotland, Europe, India, Japan, Australia, and Africa would support
this thesis as paramecia are killed by seawater. They do not form cysts
and need reasonably fresh moisture for survival, so transport to their
present locations was probably effected by continental drift. The
emergence of a single primordial endosymbiont would therefore be sufficient to populate an ancient land mass during geological times when
the continental land masses were adjacent or connected by land
bridges; then through geographical isolation the strains or races developed that we find today.

DROSOPHILA PAULISTORUM

The endosymbionts of *Paramecium aurelia* are an interesting example
of a common biological phenomenon—a relationship between two entities, one of which is transmitted by inheritance. Such transmission
occurs with intracellular parasites, symbionts, and cases of infectious
heredity. The genus *Drosophila* provides another case of the coevolution of host and an endosymbiont, in this case a microorganism that is
known as CWD (cell wall deficient) until it is more formally named.

All six of the known semispecies of *Drosophila paulistorum*—Centroamerican, Amazonian, Orinocan, Transitional, Andean–Brazilian, and Interior—have been found to harbor CWDs that seem essential to the host's welfare but cause male sterility when crossed to a different semispecies. As their names imply, these six semispecies of *D. paulistorum* are evolutionarily separated by their geographical distribution, an extrinsic isolating mechanism. Three intrinsic isolating mechanisms are also operative: sexual (behavioral or ethological) isolation, hybrid sterility, and hybrid inviability. Of these isolating mechanisms, the behavioral one is the most effective. This is seen in the few cases where the geographical distributions of two or three semispecies overlap. Where this occurs, they do not interbreed.

The genes responsible for sexual isolation are numerous: they are scattered throughout the three pairs of chromosomes and are apparently additive in their action (Ehrman, 1961, 1965; reviewed by Ehrman and Parsons, 1981). When hybrids are forced into being in the laboratory, hybrid sterility and inviability are observed. The F_1 hybrid progeny consist of fertile females and sterile males. The F_1 females will again produce fertile females and sterile male progeny when backcrossed to males of parental strains. Clearly, the isolating mechanisms prevent gamete wastage in nature.

For his final oeuvre, Dobzhansky recorded a series of phenomena that he interpreted as evidence for an incipient species, originating in the laboratory. Some years before, he had captured a gravid female of the Interior semispecies and had established an experimental population. With deference, it is perhaps best to let Dobzhansky describe what happened (Dobzhansky et al., 1976, p. 211):

A strain descended from a single female captured in the Llanos of Colombia in 1958 produced fertile hybrids of both sexes when crossed to strains of the Orinocan semispecies of *Drosophila paulistorum*. Since 1963, however, this strain, now referred to as New Llanos, gives sterile hybrid males with Orinocan strains. New Llanos gives fertile hybrids with strains of the Interior semispecies. Since this latter effect was discovered only in 1964, there is no way to ascertain whether the original Llanos would have been interfertile with Interior in 1958.

New Llanos shows little or no ethological isolation from either Orinocan or Interior semispecies, although these latter exhibit a fairly strong isolation from each other. From 1966 to 1974, artificial selection was carried on to erect an ethological isolation barrier between New Llanos and an Orinocan strain. For this purpose, two nonallelic recessive mutants were used. Homogamic matings were yielding phenotypically recognizable homozygotes, and heterogamic matings yielded wild-type heterozygotes. These latter were destroyed, while the homozygotes served as parents of subsequent generations. The selection was partially successful: strains showing a pronounced preference for homogamic matings, but not a complete ethological isolation were obtained. The experimentally induced ethological isolation is about as strong as the weakest isolation observed between semispecies of this group that occur in nature.

One possible explanation for the change in compatibility is that somehow the omnipresent CWD underwent changes induced by laboratory culturing of the host *Drosophila*.

The sterility of hybrid *D. paulistorum* is caused by a cell wall-deficient microorganism, CWD, in the testes of the hybrid males. CWD is an infectious agent that rapidly proliferates in the testes with a concomitant breakdown of spermatogenesis. This microorganism is destroyed by low pH, lipid solvents, UV, and exposure to 56°C for 30 minutes. It appears to be sensitive to tetracycline and insensitive to penicillin. CWD is also pathogenic in an unnatural host, the larva of the Mediterranean meal moth, *Ephestia kuehniella*, where it possesses killing power. It can be serially passed through *Ephestia* and retain its specificity for its original *D. paulistorum* semispecies when later reintroduced into *D. paulistorum* (Gottlieb et al., 1981).

No untreated ("uncured") *D. paulistorum*, male or female, has ever been observed to be free of CWD, and the flies do not survive long after their CWD is removed by heat shock and antibiotics. Each semispecies of *D. paulistorum* has its own highly specific endosymbiont without which it cannot thrive. CWD appears to pay its rent by providing some essential factor or service—perhaps the synthesis of vitamins. The benefits of this endosymbiont to its specific host are obvious, but the exact nature of the benefits remains in the realm of conjecture.

In the wild state, CWD is strictly inherited through the egg's cytoplasm. Laboratory-induced infections of foreign CWD produce the same fertility patterns as hybridization. Extracts of testicular CWD from one semispecies injected into a female of another semispecies will result in fertile daughters and sterile sons, showing that the hybrid genotype is not necessary for this phenomenon.

Adult females do not seem to suffer from foreign CWD, but its effects can be seen in larval inviability. Electron microscopy shows gross morphological differences between hybrid and nonhybrid embryos. When foreign CWD is carried in the egg's cytoplasm, the embryo develops abnormally and cytoplasmic blebs form (Ehrman and Daniels, 1975). Many larvae are thus lost before sexual differentiation.

The taxonomy of these CWDs has so far not been satisfactorily resolved. Ultrastructural studies reveal central fibrous networks surrounded by peripheral granulation, as described for *in vitro* cultures of *Mycoplasma hominis* (Kernaghan, 1971; Ehrman and Kernaghan, 1971). Razin (1973), in a review of mycoplasmal physiology, points out that although the internal structure and overall morphology of the *D. paulistorum* endosymbionts show "striking resemblances" to classic mycoplasmas, the presence of a duplex membrane suggests that they may be rickettsia or chlamydia.

133

Another infectious agent is a microorganism that causes the sex ratio (SR) condition in *Drosophila willistoni*. As with some other species of *Drosophila*, females of the *willistoni* group collected from natural populations will occasionally produce progeny that are entirely or almost entirely female. This condition continues in the strain indefinitely if "maintainer" males from other strains are used as fathers. The SR agent is, like CWD, maternally transmitted through the egg cytoplasm. Williamson and Poulson (1979) comprehensively reviewed what is currently known of SR in the *willistoni* group.

The SR trait was first studied in a *D. willistoni* strain from Jamaica and a *D. paulistorum* strain from Colombia (Malogolowkin and Poulson, 1957; Malogolowkin, 1958). It has since been found in *D. nebulosa* and *D. equinoxialis*; but curiously, it has not been found in *D. tropicalis*, a sibling species. Although documentation of the geographical range of the trait in each species is very poor, it is clear that the SR trait is quite widespread in the Caribbean and South America. The frequency of females with SR in a population usually ranges from 0.5 to approximately 10% (Williamson and Poulson, 1979; Marques and de Magalhaes, 1973).

One of the first indications that SR in the *willistoni* group was caused by a microorganism came from injection experiments (Malogolowkin and Poulson, 1957; Malogolowkin et al., 1959, 1960). Supernatant fractions from homogenates of SR females, when injected into non-SR strains, could produce the SR trait in recipient strains. Once established, the trait was passed on through the egg cytoplasm. The highest concentration of infective material came from adult hemolymph (Sakaguchi and Poulson, 1960, 1961). Interspecific transfer of the SR trait is possible outside the *willistoni* group, for example, to *D. melanogaster* and *D. pseudoobscura* (Williamson, 1965; Williamson and Poulson, 1979). In the natural host, death of male embryos occurs early in embryogenesis in the egg (Counce and Poulson, 1962). When transferred to other hosts, for example, *D. melanogaster*, death of males may occur later in the larval or pupal stage (Counce and Poulson, 1966).

Staining reactions and examination by light microscopy implied that the causative agent of SR was a spirochete resembling the genus *Treponema*. Electron microscopic studies, however, indicated that the organism lacked certain features of spirochetes (Williamson and Whitcomb, 1975; Williamson et al., 1977). Rather, it is morphologically identical and serologically similar to spiroplasmas (Williamson and Poulson, 1979; and for details concerning the spiroplasmas, related to mycoplasmas and CWD, see Barile and Razin, 1978).

A variety of studies indicate that there is more than one type or species of spiroplasma involved. For example, mixed infections of organisms from two species of *Drosophila* produce interference as indicated by an interruption of the SR condition in the host (Sakaguchi et al., 1965). Mixtures of hemolymphs *in vitro* can cause clumping of spiroplasmas if the hemolymphs are from two different species of *Drosophila* (Williamson and Poulson, 1979). Thus, unlike the interaction between CWD and its host, in which the endosymbiont is benign and even beneficial to the specific semispecies that carries it, different species of *Drosophila* harbor different spiroplasmas but are not fully adapted to them, because the endosymbiont continues to cause mortality of male flies.

Finally, viruses harbored by spiroplasmas vary in different species. Oishi and Poulson (1970) detected a lysogenic virus associated with *D. nebulosa* spiroplasma. When this virus is injected into a fly carrying spiroplasma derived from *D. willistoni*, the virus lyses the spiroplasma and the fly strain is "cured" (Oishi, 1971). At least six different viruses that vary in their lytic properties have been characterized (Williamson and Poulson, 1979). It appears that not only do different strains and species of flies harbor different spiroplasmas, but each type of spiroplasma has its own unique virus, to which it is tolerant. More details concerning the viruses and their morphology may be found in Williamson et al. (1977).

CONCLUSION

Endosymbiosis provides a bonanza for the scientist interested in charting coevolution. The multiple-tiered systems are especially intriguing. Research can be approached from several points of view; genetic, taxonomic, coevolutionary, ecological. Future analyses will undoubtedly elucidate the remarkable biochemistry that allows the endosymbiont to be helpful to its coevolved host and harmful to the host's near relatives. The endosymbioses that are described here—paramecium/bacteria/phage or fly/CWD/virus—exemplify *par excellence* the points made by Grun and by Margulis.

For as Grun (1976) has stated in his review of cytoplasmically inherited endosymbionts, "The evolutionary coadaptation between the cytoplasmic factors of a population of plants or animals and their nuclear genes is a specific product of the evolutionary selection pressures to which both are subject. The result is that the interaction supplies coherence to the organisms of the population." It is interesting to speculate that the dominant modes of the coevolution of endosymbionts are association, cooperation, integration, and coalescence and not "nature red in tooth and claw." To return to the thesis (Margulis, 1970) with which this chapter began—that endosymbiotic

135

processes were responsible for the origin of the eukaryotic cell—one wonders if the coevolution of these particular symbionts will progress to a further phase in the life cycle of their hosts. Like the free-living organisms that became the chloroplasts and the mitochondria of cells with which they integrated, will (have?) these endosymbionts (kappa, CWD) become organelles of the creatures they inhabit? Only time, selection, and coevolution, in concert, will tell.

CHAPTER 7

PLANT-FUNGUS
SYMBIOSES

John A. Barrett

In its original sense, symbiosis meant that two species were found in close association for a considerable part of their life cycles [De Bary: quoted by Lewis (1973a) and Cooke (1977a)]. This definition is very broad and covers a range of associations, from those in which one member is totally dependent on the other for its existence (parasitism) to those in which both symbionts cannot exist in the absence of the other (mutualism). Plants and fungi form a wide range of symbiotic associations, and this chapter will examine the ways in which these associations have evolved. Evidence of evolutionary change can be gleaned from two main sources. The first is the taxonomic relationships between different groups of organisms, from which inferences can be made about the adaptations that characterize each group. The second source is the observation of microevolutionary change at the population level.

In the first part of this chapter, I will examine the coevolutionary processes in associations between plants and parasitic fungi. Most of this evidence is derived from studies of microevolutionary changes in fungal parasites of crop plants. In the second part, I will examine the evidence for coevolution in "mutualist" associations such as lichens and mycorrhizae in which genetic evidence is lacking and inferences have to be made about the form of the reciprocal adaptation from physiological, morphological, and taxonomic studies.

PARASITIC SYMBIOSES

**Is there any evidence for heritable variation in plants
with respect to their association with parasitic fungi?**

Ever since humans have made records of their activities, the destruction of crops by disease has been a recurrent theme in these records. For example, the first recorded epidemic of ergot caused by the fungus *Claviceps purpurea* was made in AD 857 in the Rhine Valley; the role of barberry (*Berberis vulgaris*) as the alternate host of wheat rust (*Puccinia graminis*) was sufficiently well established by the seventeenth century that laws requiring the eradication of barberry were enacted in France (Horsfall and Cowling, 1978). At the beginning of and during the nineteenth century, systematic examination of cereal crop species had enabled the early plant breeders, such as Le Couteur, Cooper and Shirref, to make selections and establish improved varieties. In 1815 Knight recommended that only those selections that showed reduced symptoms of disease (resistance) should be used to establish new varieties. Among many of the examples quoted by Darwin to support his ideas on variation within populations and natural selection can be found observations of the differing susceptibility of cultivated plants to disease, e.g., "Cuthill's Black Prince (a strawberry variety) evinces a singular tendency to mildew." So from the early stages of methodical plant breeding, there are observations that lines and varieties may differ in their susceptibility to disease, and these observations imply that this variation is in part inherited (Barrett, 1981).

However, not until the "rediscovery" of Mendel's work at the beginning of the twentieth century was any progress made in the elucidation of the inheritance of disease resistance. Considering the economic and social importance of disease on crops, it is perhaps not surprising that some of the earliest papers describing Mendelian inheritance should be on disease resistance. In 1905 and 1907, Biffen published two papers that described the Mendelian basis of resistance of barley to mildew (*Erysiphe graminis*) and wheat to yellow rust (*Puccinia striiformis*). Moreover, he was able to demonstrate that "immunity is independent of any discernable morphological character, and it is practicable to breed varieties morphologically similar to one another, but immune or susceptible to attacks of certain parasitic fungi" (Biffen, 1907). This demonstration that disease resistance is inherited led to much work on the mode of inheritance of resistance to economically important diseases in most of the major crops of the world. By 1974 Day estimated that over 1000 papers had been published describing the inheritance of disease resistance. By and large, most of the papers describe resistance as being controlled by one or few genes (oligogenic resistance), but this may be an artifact, as most of the experiments have been carried out on cultivated species, and characters controlled

138

by one or few genes are easier to handle during breeding programs. However, many cases of quantitative inheritance of disease resistance (polygenic resistance) have been described. In some crops, resistance to a parasitic fungus has been demonstrated to be controlled both by single genes and polygenically (e.g., resistance to *Puccinia sorghi* in corn; Day, 1974).

Prior to intensive cultivation, each farming area had a range of "varieties" characteristic of that area. These local "varieties" (*land-races*) consisted of mixtures of different genotypes that had been maintained under mass selection by generations of growers. The early stages of methodical plant breeding involved the extraction of "improved" true-breeding lines from land-races and it is from this source that some of the resistance factors used in modern varieties were originally isolated, e.g., gene *Mlg* controlling powdery mildew (*Erysiphe graminis* f. sp. *hordei*) resistance in barley (see Barrett, 1981). However, with increasing pressure on plant breeders for even better varieties, the wild progenitors and near relatives of crop species have become increasingly exploited as sources of useful characters for incorporation into new varieties, not least as sources of disease resistance. Among the crosses in which Biffen first described the inheritance of powdery mildew resistance in barley, a resistant line of the wild barley *Hordeum spontaneum nigrum* was used. In the search for resistance to late blight of the cultivated potato caused by *Phytophthera infestans*, it was found that lines of the wild potato *Solanum demissum* showed resistance and a number of resistance genes were successfully transferred from *S. demissum* to the cultivated potato *S. tuberosum*. [See Day (1974) and Van der Plank (1963) for discussion.]

Is there evidence of heritable variation in fungal parasites with respect to their plant hosts?

During 1917 and 1918 Stakman and his colleagues published what have since become classic papers of plant pathology (Stakman et al., 1917, 1918a,b). They noted that isolates of wheat stem rust (caused by *Puccinia graminis tritici*) that were taken from different wheat varieties and used to infect a range of varieties tended to produce most disease on the varieties from which they were originally isolated. This phenomenon they termed "physiologic specialization"; although this term is not strictly biologically correct, it is still useful for describing the phenomenon. The fact that such isolates retained their specificity for the host variety was at least an indication of a genetic component of this specialization. The use of sets of varieties (differential sets) to dissect the population structure of parasite populations has become established as a basic tool of plant pathologists, and much effort has been devoted to cataloging pathogen populations using this method in most economically important diseases. Proof of the genetic basis of the pathogenicity (i.e., the ability to infect host plants) was not long in coming for Waterhouse (1929) and Newton and Brown (1930) demonstrated segregation of single loci controlling pathogenicity in stem rust [see also Johnson (1953) for review]. The elucidation of the genetic control of pathogenicity is inextricably bound to the process of resistance breeding in the host crop species. Consequently, the genetic basis of pathogenicity has arisen as a by-product of the study of the inheritance of resistance.

Is there evidence of complementary heritable variation in parasitic plant–fungus symbioses?

On the basis of observations that different genotypes of crop plants varied in their ability to resist disease-causing fungi, that this variation had a genetic basis, and that, conversely, different isolates of disease-causing fungi varied in their ability to incite disease on different "lines" of host plants, Flor began a series of experiments to investigate the genetic relationships between plants and their fungal parasites.

Flor used different varieties of flax (*Linum usitatissimum*), which could be classified as either resistant or susceptible to a defined "race" of flax rust (*Melampsora lini*), and different "races" of flax rust, which were either "virulent" (i.e., able to attack) or "avirulent" (i.e., unable to attack) on a certain variety. By crossing the different varieties of flax and testing the F_2 generation with the races to which the parental lines were susceptible, he was able to demonstrate that resistance in flax was inherited as a dominant character controlled by a single locus. Similarly, by crossing rust races that differed in their ability to infect different varieties and testing the F_2 generation on the same varieties, he showed that "virulence" in the flax rust was inherited as a recessive character controlled by a single locus (Flor, 1942, 1955, 1956). From these series of experiments Flor came to the conclusion that "for each gene conditioning rust reaction there is a specific gene conditioning pathogenicity in the parasite" (Flor, 1956). The idea of complementary genetic systems in both host and parasite first demonstrated by Flor has become known as the "gene-for-gene hypothesis." Since the publication of Flor's papers, other systems have been demonstrated to exhibit gene-for-gene relationships (see Day, 1974).

Although the "gene-for-gene hypothesis" neatly summarizes the phenomenon described by Flor, not all systems so described conform exactly to Flor's usage. Flor's original definition implied that for each locus controlling resistance in the host there is a complementary locus in the parasite controlling pathogenicity. However, current usage is to describe any system in which both resistance–susceptibility and virulence–avirulence segregate in a simple Mendelian fashion as "gene-for-gene" systems, even if the allele for "susceptibility" to a given fungus "race" is in fact an allele for resistance to another "race" of the fungus. For example, further research on the flax–flax rust system has revealed that multiple alleles at a number of loci controlling resistance exist but that the corresponding virulence factors to each of the resistance alleles are inherited at different loci, that is, in these cases resistance may be monogenic (one locus) but the complementary pathogenicity is controlled by multiple loci (Lawrence et al., 1981a,b). Notwithstanding the fact that the genetic interaction between host and parasite may not be on a strictly locus-for-locus basis, the fact remains that in many cultivated species and their fungal diseases, reciprocal adaptation appears to be controlled by a small number of genes.

The simplicity of the gene-for-gene hypothesis has, perhaps, blinded many workers to the possibility that the interaction may be more complex. A resistant variety is not resistant just because of the effects of a single locus substitution; expression of the character "resistance" is also dependent on the genetic background in which the resistance is placed. Within a group of varieties, each carrying the same major resistance factors, differences can be observed between isolates taken from each of the varieties in their ability to infect other varieties carrying the same resistance gene(s) (Wolfe and Schwarzbach, 1978; and personal communication). So although the basic "gene-for-gene" relationship remains, it is overlaid by variation due to differences in genetic background of the host plants to which the parasite can also become adapted.

Because all of the experiments on complementary genetic systems in hosts and parasites have been carried out on cultivated species, there always remains the possibility that the observation of gene-for-gene systems and the emphasis on their importance is an artifact of agriculture and plant breeding (Day, 1974). Faced with a severely diseased crop, a plant breeder will be looking for large effects in his search for resistance characters to incorporate into his breeding programs. Moreover, the screening techniques used in breeding will often only give unequivocal classification when large resistance effects are present. When varieties incorporating apparently total resistance to a parasite are exposed to a relatively heterogeneous parasite population on a large scale, it is likely that only a substantial change in the parasite will enable it to attack the new variety. It appears that in the majority of cases in fungi parasitic on crop plants the phenotypic change is outside the normal range of quantitative variation present in the parasite population, and so the change is more easily accomplished by changes at one (or few) major gene loci. This is not to say that quantitative variation has no role in the response. Once a genotype that can survive on a new host variety is present, genes that modify the effects of the major gene(s) and improve its (their) performance on that variety will be favored by natural selection. For example, the barley variety *Keg*, which carries major gene resistance to powdery mildew, was grown on an experimental plot (after it had been introduced commercially, but before it was grown on a large scale) and exposed to natural infection by powdery mildew (*Erysiphe graminis* f. sp. *hordei*). Isolates taken from the youngest leaves at the end of the growing season showed a reduced variance in pathogenicity when compared with isolates taken from older leaves that had become infected earlier in the season, when tested under standard laboratory conditions (M. S. Wolfe, personal communication).

Two further factors contribute to the demonstration of major gene resistance in most plants. First, characters controlled by simple Mendelian factors are easier to handle in breeding programs. Second, experiments on the inheritance of resistance are often carried out under controlled environmental conditions, and hence any environmental contribution to phenotypic expression will be reduced; Ellingboe (1978) has argued that by carefully adjusting environmental conditions, all host–parasite interactions can be shown to segregate in a Mendelian fashion.

By the same token, generalizations about the dominance relations of resistance–susceptibility and virulence–avirulence must be treated with caution. Following the arguments of Fisher (1928a,b, 1930), if selection is not too strong, an advantageous gene spreading through a population will acquire modifiers that enhance its expression and so such genes are more likely to be dominant than recessive. In screening breeding material for resistance, very large populations cannot be tested; therefore, only resistance genotypes present at reasonably high frequencies are likely to be detected. Because such resistance is likely to have been advantageous in the recent past of the screened population, the resistance genes detected are more likely to be more dominant than recessive. Furthermore, the breeding and screening techniques involved in the production of new varieties will themselves tend to select genetic backgrounds that enhance expression of the resistance genes. Consequently, the final commercial variety is more likely to carry resistance genes that are more dominant than recessive. When a new variety is introduced into cultivation, it is resistant because the corresponding pathogenicity is very rare, possibly because it is disadvan-

142

tageous, and in a diploid parasite, more likely to be recessive (Fisher, 1928a,b, 1930). On exposure to the new resistant variety, all genotypes except the rare recessive homozygote will die, and consequently, on the new resistant variety, the virulence gene becomes effectively fixed in one generation; there is, therefore, no possibility of heterozygotes being formed and dominance evolving (for discussion, see Sved and Mayo, 1970).

No doubt, on the basis that a simple explanation is intellectually more appealing, some host–parasite interactions have been described as exhibiting gene-for-gene relationships on the basis of single locus changes in the host plant and a phenotypic correspondence in the parasite (or *vice versa*) without any genetic tests being carried out (Rodrigues et al., 1975; Hiura, 1979). Indeed, it may well prove impossible to elucidate the genetic basis of the ability of parasites to attack hosts where the sexual stage of the fungus is not known and no parasexual cycle has been demonstrated. Even in species such as *Verticillium* spp., where sexual stages are unknown and parasexuality is known, little work has been carried out on the genetics of pathogenicity, although physiologic specialization has been demonstrated (Day, 1974; Puhalla, 1979). Thus, although there is evidence that complementary genetic systems can exist in plant host–fungal parasite systems, the fact that most of this evidence is derived from the interaction between cultivated crop plants and their parasites means that the simple genetic basis of these interactions may be an artifact.

Is there evidence of genetic change in host populations when exposed to fungal parasites?

The potato was introduced into Europe in the sixteenth century, but *Phytophthera infestans*, the causal fungus of potato late blight, was not introduced with it. Potato cultivation prospered, eventually becoming the staple food crop of the peasantry of Europe. Sometime during the 1840s *P. infestans* was introduced into Europe and gave rise to the Great Potato Blight of 1845, which led to the deaths of many thousands, perhaps millions, of people who depended on the potato. After these great epidemics, the severity of the disease declined, but it was still prevalent in western Europe. In 1833 a consignment of potatoes had been taken to Basutoland (Lesotho) and cultivated there in the absence of *P. infestans*. These potatoes were shown to be very susceptible to potato late blight when compared experimentally with early twentieth century varieties (Van der Plank, 1963). It would appear that, if the potatoes taken to Basutoland were representative of early nineteenth century European potato stocks, then these stocks were very susceptible to potato late blight and that this susceptibility may have accounted for the devastating effects of the disease when it arrived in Europe. However, the early twentieth century varieties with which they were compared were derived from postblight stocks and, because little methodical breeding in potatoes was practiced during the immediate postblight period, it seems likely that the resistance found in the later varieties was derived from resistant genotypes that had survived the Great Potato Blight.

Another American staple crop, corn, achieved worldwide distribution after the discovery and exploration of the Americas by Europeans in the fifteenth and sixteenth centuries. Corn was successfully introduced into West Africa where it was grown extensively by both commercial growers and subsistence farmers. In 1949 the rust fungus *Puccinia polysora* became established in West Africa, and the corn in that area succumbed to the disease. When lines of corn from areas where *P. polysora* was endemic were grown in West Africa alongside local lines, the local lines were more susceptible (Van der Plank, 1963; Robinson, 1976). This demonstrated that the average resistance of the crop cultivated in the absence of disease was very low; whether this was a chance effect because the lines originally introduced into West Africa carried little or no resistance or whether resistance had declined because of the absence of selection by *P. polysora* cannot be determined. Within five years, the epidemic had begun to decline as farmers planted the survivors from the previous season's epidemic and resistance slowly accumulated in the crop to a level sufficient to produce a reasonable yield.

Thus, although there is no direct evidence that fungal parasites can generate an evolutionary response in their host plant populations, there is circumstantial evidence that permits this inference to be made.

Is there evidence that evolutionary changes can occur in parasite populations in response to genetic variation in host populations?

In 1916 the North American wheat belt was devastated by an epidemic of stem rust (*Puccinia graminis* f. sp. *tritici*). As part of the reaction to this disaster, a breeding program for resistant varieties was initiated and led to the development of the first modern variety incorporating major gene resistance. However, in 1935, when a large part of the wheat-growing area of the Midwest was planted with the variety *Ceres* incorporating this resistance gene, a second epidemic hit the area. When the rust population was examined using differential sets, it was found that the population consisted predominantly of one "race" (phenotype), race 56, which was able to attack *Ceres* (Van der Plank, 1963; Barrett, 1981).

144

As agriculture has developed during the twentieth century, the following sequence of events has become more common. Drawing on the fact that genetic variation for resistance exists, plant breeders produce new varieties that are resistant to the prevailing parasite population. Because the resistant varieties produce higher yields, in part because of disease resistance, the new varieties are grown on a large scale. This increased scale of cultivation selects out of the parasite population those genotypes that can overcome the resistance; these genotypes increase in frequency, as they are the only forms of the parasite that can infect the "resistant" varieties, and "resistance breakdown" occurs. Because the benefits of resistance have now been lost, the varieties fall from favor with the growers and are usually replaced with new "resistant" varieties, and the process repeats itself. This process of producing varieties with increased resistance followed by adaptation of the parasite population and the fall from favor of the variety is known as the "boom and bust" cycle and has become more and more common wherever intensive crop husbandry is practiced, especially in cereals (for more examples, see Barrett, 1981).

The social and economic consequences of "resistance breakdown" can be very dire indeed. Perhaps one of the best documented cases of recent years is that of the southern corn leaf blight epidemic in the southern states of the United States in 1970 (Anon. 1972). Southern corn leaf blight, caused by *Helminthosporium maydis*, had long been known as an important disease of corn, and over the years breeders had gradually improved the resistance of corn varieties, by selection of polygenically controlled resistance, until the severity of the disease was of manageable proportions. Yet in 1970 an epidemic caused substantial damage; statewide yield losses of 50% were not uncommon. The cause of the epidemic was a new race of *H. maydis*, which was able to attack most of the cultivars in use at the time. At first sight this was surprising because most commercial cultivars of corn are either single- or double-cross hybrids and genotypically fairly heterogeneous. However, to improve the efficiency of the crossing programs that give the commercial hybrids, cytoplasmic factors that confer male sterility are used. The one common factor in all of the varieties that had been attacked by *H. maydis* during this epidemic was that they all possessed the same cytoplasmic male-sterility factor, Texas male-sterile cytoplasm (*Tms*). It has been estimated that at the time of the 1970 epidemic approximately 10^{15} cytoplasmically identical plants were growing in the United States (Browning, 1972). It was possession of this cytoplasmic factor that made the corn plants susceptible to a toxin produced by a new race of *H. maydis*, race *T*. After the epidemic had receded and the postmortem on its cause was carried out, it was found that in a preserved collection of *H. maydis* samples, from well before the epidemic, genotypes were present that could attack cultivars carrying *Tms* cytoplasm. The existence of these races had been noted at the time of introduction of the *Tms*-carrying varieties, but these races were predominantly of a mating type different from that that eventually caused the great epidemic of 1970 and did not appear to have the ability to attack the corn plants very severely. However, the selection imposed by the extensive use of these varieties shifted the population of *H. maydis* toward increased ability to attack these varieties.

Pearl millet is a staple crop of subsistence farmers on poor land in India. In order to increase production, hybrid varieties were introduced and total yields more than doubled over a period of 20 years. However, in 1971 the pearl millet crop was devastated by an epidemic of downy mildew (*Sclerospora graminicola*). On investigation it was found that this epidemic was a repeat of the southern corn leaf blight–*Tms* story. In most of the varieties that succumbed to the disease, just one source of cytoplasmic male sterility (Tift 23A) had been used (Safeeulla, 1977). So, changes in the genetic composition of crop populations can induce evolutionary responses in parasite populations that can have disastrous consequences.

VARIATION AND EVOLUTION IN NATURAL ECOSYSTEMS

The constant breakdown of resistant varieties in agriculture has forced plant breeders to continuously screen for resistance genes in wild populations of the progenitors and near relatives of cultivated species. So, to a large extent the study of wild populations has been to identify and conserve resistance genes and other useful characters as sources for plant breeding. However, a few studies have been devoted to a close examination of the variation present in natural populations; little or no work has been carried out to explore the genetic interactions in plant host–fungal parasite systems unrelated to crop species or to follow the dynamics of such systems under genuinely "wild" conditions. The most extensive studies of this type have been carried out on the wild barley *Hordeum spontaneum* and its fungal parasite *Erysiphe graminis* and on the wild oat species *Avena sterilis* and *Avena barbata* and their fungal parasite *Puccinia coronata* (crown rust)in Israel (Dinoor, 1974, 1977; Eshed and Dinoor, 1981; Fischbeck et al., 1976; Segal et al., 1980; Wahl, 1970; Wahl et al., 1978). By testing plants grown from seed collected in the wild with known genotypes of the fungal parasites, it is possible to detect the presence of major resistance genes. By testing isolates of the fungus on a range of different varieties carrying different resistance genes, it is possible to

146